Interactions in
Multiagent
Systems

Interactions in
Multiagent
Systems

Jianye Hao
Tianjin University, China

Ho-Fung Leung
The Chinese University of Hong Kong, China

World Scientific

NEW JERSEY · LONDON · SINGAPORE · BEIJING · SHANGHAI · HONG KONG · TAIPEI · CHENNAI · TOKYO

Published by

World Scientific Publishing Co. Pte. Ltd.

5 Toh Tuck Link, Singapore 596224

USA office: 27 Warren Street, Suite 401-402, Hackensack, NJ 07601

UK office: 57 Shelton Street, Covent Garden, London WC2H 9HE

Library of Congress Cataloging-in-Publication Data

Names: Hao, Jianye, author. | Leung, Ho-Fung, author.

Title: Interactions in multiagent systems / Jianye Hao (Tianjin University, China),
    Ho-Fung Leung (The Chinese University of Hong Kong, China).

Description: New Jersey : World Scientific, [2018] | Includes bibliographical references and index.

Identifiers: LCCN 2018008585 | ISBN 9789813208735 (hc)

Subjects: LCSH: Multiagent systems.

Classification: LCC QA76.76.I58 H37 2018 | DDC 006.3/0285436--dc23

LC record available at https://lccn.loc.gov/2018008585

British Library Cataloguing-in-Publication Data

A catalogue record for this book is available from the British Library.

For any available supplementary material, please visit
http://www.worldscientific.com/worldscibooks/10.1142/10414#t=suppl

Desk Editor: Herbert Moses

Typeset by Stallion Press
Email: enquiries@stallionpress.com

Printed in Singapore

Foreword

Autonomous Agents and Multiagent Systems (AAMAS) has been one of the research hotspots in artificial intelligence for the last 20 years. I have participated in research in this area since the millennium and have witnessed its healthy growth. The pioneering works of researchers in the early days, most of whom were from Europe, Oceania, and North America, has laid a solid foundation for this research area. In recent years, a growing number of young researchers from the Asia-Pacific region have recognized the significance of this research topic, and have thus joined to strengthen this prosperous research community.

I am happy to see the publication of this book edited by Jianye Hao and Ho-Fung Leung. Ho-Fung is an old friend of mine, who is a prolific researcher in AAMAS and related areas. Jianye is an excellent young researcher in AAMAS. His research work and publications during the past a few years in the research of AAMAS have been impressive. In this edited volume, they have solicited works from many outstanding young researchers from the Asia-Pacific region who represent the vital new force of AAMAS research. While the main theme of this book is the interactions in multiagent systems, it covers a wide range of topics including automated negotiation, coordination, diffusion convergence of collective behaviors, multiagent reinforcement learning, norm emergence, planning, POMDP, reputation, and task allocation. This book is thus a timely summary of many important new results.

It is my pleasure to introduce this book to you, which sketches a profile of the research work of AAMAS researchers in the Asia-Pacific.

Zhongzhi Shi
Key Laboratory of Intelligent Information Processing
Institute of Computing Technology, Chinese Academy of Sciences
China

About the Editors

 **Jianye Hao**（郝建业）is currently an Associate Professor in School of Computer Software at Tianjin University. Before that, he worked as a postdoctoral fellow at MIT and SUTD from 2013 to 2015. He received his Ph.D. degree in computer science and engineering from The Chinese University of Hong Kong in 2013 and B.E. degree from School of Computer Science and Technology at Harbin Institute of Technology in 2008. His research interests include multi-agent systems, (deep) reinforcement learning and game theory. He has published over 50 research papers in leading conferences (e.g., *AAMAS, IJCAI, AAAI, FSE*) and journals (e.g., *JAAMAS* and *ACM/IEEE Trans*).

Ho-Fung Leung (梁浩锋) is a Professor in the Department of Computer Science and Engineering at The Chinese University of Hong Kong. He receives his B.Sc. and M.Phil degrees in computer science from The Chinese University of Hong Kong, and his Ph.D. degree from University of London with DIC (Diploma of Imperial College) in computing from Imperial College London. His research interests include ontologies, intelligent agents, and complex systems.

About the Contributors

Yujing Hu (胡裕靖) received his Bachelor's degree in computer science and technology from the Department of Computer Science and Technology, Nanjing University, Nanjing, China, in 2010, where he also received the Ph.D. degree in computer application technology. His research interests include reinforcement learning, multiagent learning and game theory.

Hao Wang (王皓)、received his Bachelor's degree in mathematics from the Department of Mathematics, Nanjing University (NJU) in 2005 and Master's Degree in computer science from the Department of Computer Science and Technology, NJU in 2008. He received his Ph.D. from the Department of Computer Science at The University of Hong Kong (HKU) in 2014. His research interests include data indexing, recommender system, reinforcement learning and transfer learning.

Yunkai Zhuang (庄辊恺) received his Bachelor's degree in digital media technology from the College of Digital Media, Jiangnan University, Wuxi, China, in 2014. He is currently pursuing the Ph.D. degree in computer application technology in the Department of Computer Science and Technology, Nanjing University, Nanjing, China. His research interests include reinforcement learning, multiagent systems and game theory.

Chao Yu (余超) received his Ph.D. degree in computer science from the University of Wollongong, Australia, in 2014. Now, he is an Associated Professor in the School of Computer Science and Technology at the Dalian University of Technology, Dalian, China. His research interests include multi-agent systems and reinforcement learning.

Hongtao Lv (吕洪涛) received his B.S. degree in Computer Science and Technology from the Dalian University of Technology in 2017. He is currently a Ph.D. candidate at the Department of Computer Science and Engineering in Shanghai Jiao Tong University. His research interests include mechanism design, artificial intelligence, algorithmic game theory and its application.

Hongwei Ge (葛宏伟) received B.S. and M.S. degrees in mathematics from Jilin University, China, and the Ph.D. degree in computer application technology from Jilin University, in 2006. He is currently an Associate Professor and a Vice Dean in the School of Computer Science and Technology, Dalian University of Technology, Dalian, China. His research interests are machine learning, computational intelligence, optimization and modeling, transfer learning, and deep learning.

Liang Sun (孙亮) received his B.E. degree in computer science and technology from Xidian University, Xi'an, China, and the M.S. degree in computer application technology from Jilin University, Changchun, China, in 2003 and 2006, respectively. He received double Ph.D. degree from Kochi University and Jilin University in March, 2012 and June 2012, respectively. He is currently with the School of Computer Science and Technology, Dalian University of Technology, Dalian, China. His main research interests lie in computational intelligence and machine learning methods.

Jun Meng (孟军) received her Ph.D. degree from the Dalian University of Technology in 2012. She is currently a Professor in the School of Computer Science and Technology at Dalian University of Technology. She has published more than 50 research papers in various journals and conferences. Her research interests include artificial intelligence, data mining and machine learning.

Bingcai Chen (陈炳才) received his M.S and Ph.D. degrees in information and communication engineering from Harbin Institute of Technology, Harbin, China, in 2003 and 2007, respectively. He has been a visiting scholar in University of British Columbia, Canada, in 2015. He is now an Associate Professor in the School of Computer Science and Technology at Dalian University of Technology, Dalian, China, and in the School of Computer Science and Technology, Xinjiang Normal University, Urumqi, China. His current research interests include multiagent systems, machine learning and computer vision, etc.

Han Yu (于涵) is a Nanyang Assistant Professor (NAP) at the School of Computer Science and Engineering (SCSE), Nanyang Technological University (NTU), Singapore. He received his B.E. (Hons) degree and Ph.D. degree from SCSE, NTU in 2007 and 2014, respectively. He held the prestigious Lee Kuan Yew Post-Doctoral Fellowship (LKY PDF) from 2015 to 2018, and the Singapore Millennium Foundation (SMF) Ph.D. Scholarship from 2008 to 2012. His research focuses on stochastic optimization for data-driven algorithmic management in large-scale collaboration systems. Specifically, his work seeks to develop advanced AI techniques to proactively motivate, organise and manage crowdsourcing workers so as to achieve efficient utilisation of their intelligence, expertise, time and resources. He has published over 95 research papers in leading international conferences and journals including *AAAI, IJCAI, ASE, AAMAS, Proceedings of the IEEE*, as well as the *Nature Research* Journals — *NPJ Science of Learning, Scientific Data* and *Scientific Reports*. His research work has won 10 conference and journal awards (including from AAAI and IJCAI). In both 2014 and 2015, he was selected to be one of the Top 10 Global Young Scientists under 35 during the Global Young Scientists Summit (GYSS).

Chunyan Miao (苗春燕) received her B.S. degree from Shandong University, Jinan, China, in 1988, and the M.S. and Ph.D. degrees in Nanyang Technological University, Singapore, in 1998 and 2003, respectively. She is currently a Professor in the School of Computer Science and Engineering at Nanyang Technological University (NTU), and the Director of the Joint NTU-UBC Research Centre of Excellence in Active Living for the Elderly (LILY). Her research focus is on infusing intelligent agents into interactive new media (virtual, mixed, mobile and pervasive media) to create novel experiences and dimensions in game design, interactive narrative and other real-world agent systems.

Bo An (安波) is an Associate Professor at the School of Computer Science and Engineering of the Nanyang Technological University. Prior to joining NTU in 2013, he spent 1 year as an Associate Professor at the Institute of Computing Technology of the Chinese Academy of Sciences. During October 2010 to June 2012, he was a Postdoctoral Researcher at the University of Southern California, working with Professor Milind Tambe. He received the Ph.D. degree in Computer Science from the University of Massachusetts, Amherst, where he was advised by Professor Victor Lesser. Prior to that, he received the B.Sc. and M.Sc. degrees in computer science from Chongqing University, China. His research interests include artificial intelligence, multi-agent systems, game theory, automated negotiation, resource allocation, and optimization. He has published over 70 referred papers at top conferences (AAMAS, IJCAI, AAAI,

ICAPS, and KDD) and journals (*JAAMAS, AIJ* and *ACM/IEEE Transactions*). His research results in complex resource allocation have been applied to e-commerce, sensor networks, distributed streaming processing systems and cloud computing. His work on applying game theory to security has been applied to develop game-theoretic randomization software that is currently deployed by the United States Federal Air Marshals Service, the United States Coast Guard and wildlife conservation organizations.

Zhiqi Shen (申志奇) is a Senior Research Scientist with the School of Computer Science and Engineering (SCSE), Nanyang Technological University, Singapore. He obtained B.Sc. in computer science and technology from Peking University, M.E. in computer engineering in Beijing University of Technology, and Ph.D. in Nanyang Technological University, respectively. His research interests include multi-agent systems (MAS), goal oriented modeling, agent augmented interactive media, and interactive storytelling.

Cyril Leung received his B.Sc. (honors) degree from Imperial College, University of London, UK, and the M.S. and Ph.D. degrees in electrical engineering from Stanford University. He has been an Assistant Professor in the Department of Electrical Engineering and Computer Science, Massachusetts Institute of Technology and the Department of Systems Engineering and Computing Science, Carleton University. He is currently a Professor in the Department of Electrical and Computer Engineering, UBC. He served as Associate Dean, Research and Graduate Studies, in the Faculty of Applied Science from 2008 to 2011. His current research interests are in wireless communications systems as well as technologies to support active independent living for the elderly.

Zongzhang Zhang (章宗长) is an Associate Professor in School of Computer Science and Technology at Soochow University. He received a Ph.D. in computer science from University of Science and Technology of China in 2012. He worked as a research fellow at National University of Singapore from 2012 to 2014 and as a visiting scholar at Rutgers University from 2010 to 2011. His research directions include planning under uncertainty, reinforcement learning and multi-agent systems. He has authored or co-authored over 20 papers in leading international conferences and prestigious journals, including *IJCAI, AAAI, ICML, UAI, ICAPS, AAMAS*, and *Frontiers of Computer Science*.

Mykel Kochenderfer is an Assistant Professor of Aeronautics and Astronautics at Stanford University. Prof. Kochenderfer is the director of the Stanford Intelligent Systems Laboratory (SISL), conducting research on advanced algorithms and analytical methods for the design of robust decision-making systems. Prior to joining the faculty, he was at MIT Lincoln Laboratory where he worked on airspace modeling and aircraft collision avoidance, with his early work leading to the establishment of the ACAS X program. He received a Ph.D. from the University of Edinburgh and B.S. and M.S. degrees in computer science from Stanford University. He is the author of *Decision Making under Uncertainty: Theory and Application* from MIT Press. He is a third generation pilot.

Chengwei Zhang（张程伟）received his B.S. degree in pure and applied mathematics from Northwestern Polytechnical University in 2011. Currently, he is a Ph.D. candidate in technology of computer software and theory from Tianjin University. His research interests include reinforcement learning, multiagent system and nonlinear dynamic theories.

Xiaohong Li（李晓红）is a full tenured Professor of the School of Computer Science and Technology, Tianjin University, Tianjin, China. Her current research interests include knowledge engineering, trusted computing and security software engineering.

Zhiyong Feng（冯志勇）is a Professor and Associate Director of the School of Computer Software, Tianjin University. His research interests lie in affective computing, cloud computing, HCI, pervasive computing and information security. He is a member of the IEEE and a member of the board of supervisors of China Computer Federation.

Wanli Xue（薛万利）received his B.S. degree in pure and applied mathematics from Tianjin Polytechnic University in 2009 and M.S. degree in Software Engineering from Tianjin University in 2013. Currently, he is a Ph.D. candidate in Technology of Computer Application from Tianjin University. His research interests include visual tracking and machine learning.

Jun Wu (吴骏) is a Research Fellow of the Department of Computer Science and Technology, Nanjing University, and a staff of the National Key Laboratory for Novel Software Technology. He received his Ph.D. degree in computer science from Nanjing University in 2011. His research interests are in autonomous agent and multiagent systems, game theory and algorithmic mechanism design.

Lei Zhang (张雷) is a Research Fellow of the Department of Computer Science and Technology, Nanjing University. He received Ph.D. (2014) and B.S. (2008) degrees in computer science from Nanjing University. He is also a staff of the National Key Laboratory for Novel Software Technology. His research interests include artificial intelligence (multiagent systems) and economic theory (game theory and mechanism design).

Yu Qiao (乔羽) is currently a second year Ph.D. student of Department of Computer Science and Technology in Nanjing University and a member of IIP Group, led by Prof. Chongjun Wang. He received his Bachelor's degree in computer science in 2016 from Xi'an Jiaotong University, Xi'an China. His research interests are in autonomous agent and multiagent systems, game theory and algorithmic mechanism design.

Chongjun Wang （王崇骏） is a Professor at the Dept. of Computer Science and Technology, Nanjing University, a staff of the National Key Laboratory for Novel Software Technology, and the director of the IIP Group, Nanjing University. He received his Ph.D. degree in computer science from Nanjing University in 2004. Currently, his research interests include autonomous agent and multiagent systems, complex network analysis, big data analysis and intelligent systems. He published two books and over 50 papers in journals and proceedings in recent years. He has won research funding from several competitive sources such as the National Key Research and Development Program of China, the National Natural Science Foundation of China and the National 973 Program of China.

Siqi Chen （陈锶奇） obtained his Master and Ph.D. degrees (in 2013 and 2014, respectively) at Maastricht University under supervision of Prof. Gerhard Weiss and Prof. Karl Tuyls. Then he was a postdoc research fellow at Department of Knowledge Engineering (DKE), Maastricht University for another year. He is now an Associate Professor at College of Computer and Information Science, Southwest University. His main research interests are in distributed artificial intelligence, machine learning, game theory, multiagent systems and adaptive and autonomous systems.

Gerhard Weiss is a Full Professor of Artificial Intelligence and Computer Science at the Department of Data Science and Knowledge Engineering (DKE) at Maastricht University since 2009. He is the chair of DKE, and within DKE he is head of the "Robots, Agents, and Interaction" (RAI) research group and co-director of the Swarmlab robotics laboratory. Before joining Maastricht University, he was the Scientific Director of SCCH GmbH (Austria), where his main responsibilities were e.g. the supervision and control of R&D projects for companies, research and innovation management. His main research interests are in automated knowledge processing and the foundations and practical applications of artificial intelligence and multiagent systems. His primary research focus is on machine learning, with applications in various domains such as automated negotiation, scheduling, manufacturing, brain–computer interfaces and cognitive robotics.

Tianpei Yang（杨天培）is currently a Ph.D. candidate in School of Computer Software from Tianjin University. She received her M.S. degree in software engineering from Tianjin University in 2017 and B.E. degree from School of Computer Software at Tianjin University in 2014. Her research interests include artificial intelligence, multi-agent systems and reinforcement learning.

Zhaopeng Meng （孟昭鹏）is currently a Professor in School of Computer Software at Tianjin University and Vice President of Tianjin University of Traditional Chinese Medicine. He received his Ph.D. degree in Computer Science and Technology from Tianjin University. His research interests include Big Data Computing, Internet of Things Software and Systems, Computer Vision and Human–Computer Interaction.

Zan Wang （王赞）received his Ph.D. degree in Software Engineering from Tianjin University. Currently, he is an Associate Professor in School of Computer Software from Tianjin University. His research interests include Software Testing, Software Analysis and Machine Learning.

Yichuan Jiang (蒋嶷川) received his Ph.D. degree in computer science from Fudan University, Shanghai, China, in 2005. He is currently a Full Professor with the Distributed Intelligence and Social Computing Laboratory, School of Computer Science and Engineering, Southeast University, Nanjing, China. His main research interests include multiagent systems, social networks and social computing. He has published more than 80 scientific articles in refereed journals and conference proceedings. He won the best paper award and best student paper award, respectively, from PRIMA and ICTAI. He is a senior member of IEEE.

Yifeng Zhou (周一峰) received his Ph.D. degree in learning science from Southeast University, Nanjing, China, in 2016. He is currently an Assistant Professor with the Distributed Intelligence and Social Computing Laboratory, School of Computer Science and Engineering, Southeast University, Nanjing, China. His main research interests include multiagent systems and social networks. He has published several scientific articles in refereed journals and conference proceedings, such as the *IEEE Transactions on Parallel and Distributed Systems, ACM Transactions on Autonomous and Adaptive Systems, Chaos, Solitons & Fractals, ICTAI* and *IAT*.

Fuhan Yan (严富函) received his B.S. degree and M.S. degree at the School of Computer Science and Engineering, Southeast University, Nanjing, China, in 2013 and 2015, respectively. He is currently pursuing the Ph.D. degree at the Distributed Intelligence and Social Computing Laboratory, School of Computer Science and Engineering, Southeast University, Nanjing, China. His current research interest is multiagent systems. He has published several scientific articles in refereed journals and conference proceedings, such as the *European Conference on Artificial Intelligence (ECAI), International Conference on Autonomous Agents and Multiagent Systems (AAMAS)* and *Physica A: Statistical Mechanics and its Applications*.

Yunpeng Li (李云鹏) received his B.S. degree at the School of Computer Science and Engineering, Southeast University, Nanjing, China, in 2013. He is currently pursuing the Ph.D. degree at the Distributed Intelligence and Social Computing Laboratory, School of Computer Science and Engineering, Southeast University, Nanjing, China. His current research interests include multiagent systems, fair division and services computing. He has published several scientific articles in refereed journals and conference proceedings, such as the *ACM Transactions on Autonomous and Adaptive Systems, the International Conference on Autonomous Agents and Multiagent Systems (AAMAS)* and the *IEEE International Conference on Web Services (ICWS)*.

Yingke Chen (陈盈科) is a Lecture at College of Computer Science in Sichuan University, China. He received his Ph.D. in 2013 from Aalborg University, Denmark. Dr. Chen was a post-doctoral research associate in Queen's University Belfast, UK, and Georgia University, USA, respectively, before joining Sichuan University. His current research interests include intelligent agents, decision making and their applications in autonomous systems.

Prashant Doshi is a Professor of Computer Science at The University of Georgia, USA, and founding director of the Faculty of Robotics at UGA (an OVPR initiative). His research interests lie in artificial intelligence and robotics, and specifically in decision making under uncertainty in multiagent settings, game theory and robot learning. He was a visiting professor at the University of Waterloo in 2015. He has published over 80 papers in journals, conferences and other forums in the fields of agents and AI. His research has led to publications in the *Journal of AI Research, AAMAS, AAAI,* and *IJCAI conferences* among others.

Yinghui Pan (潘颖慧) is an Associate Professor at Department of Information Management in Jiangxi University of Finance and Economics, China. She received her Ph.D. in 2012 from Xiamen University, China. Her current research interests include intelligent agents and decision making. Most of her publications appear in *Journal of Approximate Reasoning, Agent and Data Mining Interaction, AAMAS* and *IEEE/WIC/ACM International Conference on Intelligent Agent Technology.*

Jing Tang (唐静) received her Ph.D. degree in computational intelligence from Nanyang Technological University, Singapore, in 2007. She is a Lecturer with the School of Computing, Teesside University, Middlesbrough, UK. Her research interests include computational intelligence and decision making for computer games and big data analytics. Most of her research articles appear in mainstream evolutionary computation conferences, such as *IEEE Congress on Evolutionary Computation and the Genetic and Evolutionary Computation Conference.*

Contents

3.   Making Efficient Reputation-Aware Decisions in Multiagent Systems   43

*Han Yu, Chunyan Miao, Bo An, Zhiqi Shen
and Cyril Leung*

4.   Decision-Theoretic Planning in Partially Observable Environments   65

Zongzhang Zhang and Mykel Kochenderfer

Chengwei Zhang, Xiaohong Li, Zhiyong Feng and Wanli Xue

6. Task Allocation in Multiagent Systems: A Survey of Some Interesting Aspects 107

Jun Wu, Lei Zhang, Yu Qiao and Chongjun Wang

9. Diffusion Convergence in the Collective Interactions of Large-scale Multiagent Systems

*Yichuan Jiang, Yifeng Zhou, Fuhan Yan
and Yunpeng Li*

**10. Incorporating Inference into Online Planning
in Multiagent Settings** **229**

*Yingke Chen, Prashant Doshi, Jing Tang
and Yinghui Pan*

Chapter 1

Scalability of Multiagent Reinforcement Learning

Yunkai Zhuang, Yujing Hu and Hao Wang*

*State Key Laboratory for Novel Software Technology,
Collaborative Innovation Center of Novel Software Technology
and Industrialization, Nanjing University, Nanjing 210008, China*

*wanghao.hku@gmail.com

Equilibrium-based multiagent reinforcement learning (MARL) is an important approach when deal with cooperative and competitive problems in multiagent systems. However, most of the equilibrium-based MARL algorithms cannot scale due to the high time complexity which arises for a large number of computationally expensive equilibrium computations. This problem can be solved from two aspects: 1) simplify equilibrium calculation and 2) reduce the number of equilibrium calculations. From these two points, this chapter will introduce three kinds of algorithms which can improve the scalability of MARL significantly.

1.1 Introduction

In spite of the rapid development of multiagent reinforcement learning (MARL) theories and algorithms (Wunder *et al.*, 2010; Claus and Boutilier, 1998; Kok *et al.*, 2004; Spaan and Melo, 2008), there are still many problems when compared with other multiagent system (MAS) techniques due to some limitations of the existing MARL methods. With the increase of the agent's number, the state space induced exponential growth. In practical applications, MARL

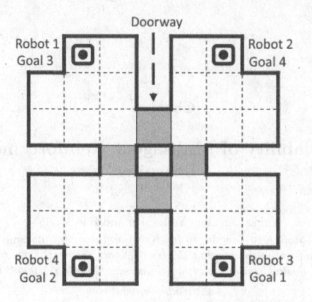

Fig. 1.1. A simple example of MAS with sparse interactions.

relies on the tightly coupled learning process. The calculation of equilibrium (e.g., Nash equilibrium) for each time step and each joint state needs vast computing resources.

Fortunately, in most of the MAS, the interaction between agents is always sparse. As we can see in Fig. 1.1, Robot 1 and Robot 3 can act independently most of the time. But when they move near the door-way, collision is more likely to occur, which leads to much more interactions between the two agents. Taking advantage of this property, many excellent algorithms are proposed in order to scale the MAS larger and larger. The main idea of these methods is to divide the process into two parts. Each agent in the system takes the joint policy only when it has detected other agents have notable affections on it. Otherwise, agent acts independently. We will introduce some of the most representative algorithms that deal with the scalability of MARL. In the following section, we review some basic concepts of MARL and game theory.

Definition 1.1. An n-agent ($n \geq 2$) Markov Decision Process is a tuple $\langle N, S, \{A_i\}_{i=1}^n, \{R_i\}_{i=1}^n, T \rangle$, where N is the set of agents, S is the state space of the environment, A_i is the action space of agent

i $(i = 1, \ldots, n)$. $A = A_1 \times \cdots \times A_n$ is the joint action space. R_i : $S \times A \times S \to [0, 1]$ is the transition function.

On this occasion, the joint policy of the whole MAS is denoted by $\pi = (\pi_1, \ldots, \pi_n)$, where π_i is the policy of agent i, $\pi_i : S \times A_i \to [0, 1]$. The state-value function V_i^π gives expected values for agent i under joint policy π at every state s

$$V_i^\pi(s) = E_\pi \left\{ \sum_{k=0}^\infty \gamma^k r_i^{t+k} | s_t = s \right\} \tag{1.1}$$

Thus, we can define the state–action value function of agent i under policy π as

$$Q_i^\pi(s, \vec{a}) = E_\pi \left\{ \sum_{k=0}^\infty \gamma^k r_i^{t+k} | s_t = s, \vec{a}_t = \vec{a} \right\} \tag{1.2}$$

where $\vec{a} \in A$ denotes a joint action and r_i^{t+k} is agent i's reward at time step $(t + k)$.

1.2 Coordinating Q-Learning

Coordinating Q-learning (CQ-learning) (De Hauwere *et al.*, 2010) is a popular method to solve multiagent problems. Aiming at reducing agent's state space, CQ-learning starts from single agent representation. Then, the algorithm identifies states in which the agent should take other agents into consideration. Finally, the agent will choose its preferred action and condition the need for coordination.

Assuming each agent has already learnt an optimal single agent policy in the environment, they have a model of its expected rewards for every state–action pair. We can use Student's *t*-test to detect if there are changes in the observed rewards for the selected state–action pair. There may be two kinds of observations:

(1) The algorithm will detect changes in the rewards the agent receives. If there is a change in the received immediate reward, the algorithm will mark this state. After that, the algorithm collects new samples from joint state space and finds out whether collisions occur or not. If collision occurs, this state is marked as being *dangerous* and the agent's state space is augmented by

adding this joint state. Otherwise, the state is marked as being *safe* and the agent's action in this state is independent of the states of other agents.

(2) The algorithm detects the changes in the rewards the agent receives and the rewards are the same as one distribution. We can say that there is no interaction between the agents. In this case, the agent can continue to act as if it was alone.

When an agent arrives at a marked state, there will be a judgment of whether it is in one of the joint states where it must take the other agent into consideration. If so, an action is selected using the following update rule:

$$Q_k^j(js, a_k) \leftarrow (1 - \alpha_t)Q_k^j(js, a_k) + \alpha_t[r(js, a_k) + \gamma \max_{a'_k} Q(s', a'_k)]$$

(1.3)

where Q_k is the Q-value of the independent states, and Q_k^j is the joint states (js). The Q-value of joint states is initialized arbitrarily, while the Q-table of single states of an agent is used to bootstrap the Q-values of the states that were argued to joint states. The algorithm is formally described in Algorithm 1.1.

CQ-learning starts with independently trained agents and maintains statistics on observed returns. The independent state space of an agent is expanded with relevant state information about other agents in order to learn a better policy when the statistics indicate a change in the policy of single agent is needed. This is the key idea of sparse interactive MARL.

1.3 Negotiation-based MARL

Equilibrium-based MARL is another important approach in MARL, which combines reinforcement learning and game theory. It takes Markov games as the framework and learns equilibrium policies by introducing various equilibrium solution concepts in game theory.

The general framework of equilibrium-based MARL is shown in Algorithm 1.2.

Algorithm 1.1: CQ-learning algorithm for agent k

1:     Initialize Q_k and Q_k^j
2:     **while true do**
3:       **if** $\forall$ Agents k, state s_k of Agent k is a safe state **then**
4:         Select a_k for Agent k from Q_k
5:       **else**
6:         Select a_k for Agent k from Q_k^j
7:       **end if**
8:       $\forall$ Agents A_k, sample $\langle s_k, a_k, rk \rangle$
9:       **if** t-test detects difference in observed rewards vs expected rewards **then**
10:        **for** $\forall$ seen states s_k of agent A_k **do**
11:          **if** t-test detects difference between independent state s and joint state js **then**
12:            add js to Q_k^j
13:            mark js as dangerous
14:          **else**
15:            mark js as safe
16:          **end if**
17:        **end for**
18:       **end if**
19:       **if** s_k is safe for Agent k **then**
20:         No need to update $Q_k(s)$
21:       **else**
22:         Update $Q_k^j(js, a_k) \leftarrow (1 - \alpha_t)Q_k^j(js, a_k)$ $+\alpha_t \left[r(js, a_k) + \gamma \max_{a_k'} Q(s', a_k') \right]$
23:       **end if**
24:    **end while**

Most existing equilibrium-based algorithms involve computing mixed strategy equilibria and require agents to replicate the other agents' value functions for equilibrium computing, which is not realistic in many practical conditions. Agents are unwilling to share their value functions due to privacy or safety concerns in some cases. To deal with this problem, negotiation-based Q-learning Hu *et al.* (2015c) is proposed. Given that mixed strategy equilibrium is often computationally expensive, pure strategy equilibrium solution

---

Algorithm 1.2: The general framework of game theory-based MARL

Input: Learning rate α, discount factor γ, exploration factorϵ

1:　　Initialization. $\forall s \in S, \forall i \in N, \forall \vec{a}, Q_i(s, \vec{a}) \leftarrow 0$

2:　　**for each** episode **do**

3:　　　Initialize state s

4:　　　**repeat**

5:　　　$\vec{a} \leftarrow \Omega(Q_1(s), \ldots, Q_n(s))$with ϵ-greedy policy
　　　　　/*Ω is for computing an equilibrium*/

6:　　　　**for each** *agent* $i{\in}N$ **do**

7:　　　　　Receive the experience $(s, \vec{a}, r_i, s'$

8:　　　　　$Q_i(s, \vec{a}) \leftarrow (1 - \alpha)Q_i(s, \vec{a}) + \alpha(r_i + \gamma\Phi_i(s'))$
　/*　　Φ_i is the expected value of the equilibrium in state s' for
　agent i*/

10:　　　$s \leftarrow s'$

11:　　**until** s is a terminal state

---

concept is adopted. Three types of pure strategy profiles can

be utilized as equilibrium solution concepts: pure strategy Nash equilibrium (PNE), equilibrium-dominating strategy profile (EDSP) and non-strict EDSP.

Since we focus on pure strategy equilibria, one obvious choice is PNE. In PNE, each agent's action is the best response to the other agents' joint action. But sometimes there may be no PNE in a game. What's more, it may be Pareto dominated by a strategy profile which is not a PNE. Take the prisoners' dilemma (PD) as an example. The PNE of this particular game is (Confess, Confess). However, if agents choose (Deny, Deny), which is a Pareto optimal solution, both of them will obtain a higher payoff. Since NE is based on the assumption of full rationality, if agents only know their own utilities, (Deny, Deny) must be more attractive than (Confess, Confess) to both of them. EDSP is a strategy profile in which Pareto dominates an NE.

(A, B)	Confess	Deny
Confess	$(-9, -9)$	$(0, -10)$
Deny	$(-10, 0)$	$(-1, -1)$

Definition 1.2. In an n-agent $(n \geq 2)$ normal-form game Γ, a joint action $\vec{a} \in A$ is an EDSP if there exists a PNE $\vec{e} \in A$ such that

$$U_i(\vec{a}) \geq U_i(\vec{e}), \quad i = 1, 2, \ldots, n \tag{1.4}$$

Definition 1.3. In an n-agent $(n \geq 2)$ normal-form game Γ, a joint action $\vec{a} \in A$ is a non-strict EDSP if $\forall i \in N$, there exists a PNE $(\vec{e_i}) \in A$ such that

$$U_i(\vec{a}) \geq U_i(\vec{e_i}), \quad i = 1, 2, \ldots, n \tag{1.5}$$

The index i of $\vec{e_i}$ indicates this equilibrium is for agent i and may be different from those of the other agents. Obviously, an EDSP is a special case of non-strict EDSP. The three types of strategy profiles can be further proved to be symmetric meta equilibria.

A multistep negotiation process is provided to find the three strategy profiles without disclosing each agent's value function in MARL. Each agent asks other agents whether a joint action is acceptable or not, while each agent answers similar questions received from the other agents at the same time. Then a preference matrix can be obtained.

Negotiation for the Set of Pure Strategy Nash Equilibria: The first step is to find the set of strategy profiles that are potentially Nash equilibria according to agent i's own utilities. Then agent i's strategy profiles are filtered out from it's J_{NE} by asking other agents whether these strategy profiles are also in their J_{NE} sets.

Negotiation for the Set of Non-strict EDSP: Each agent first computes potential non-strict EDSPs, then obtains the intersection through negotiation.

The three identified strategy profiles can be proved to be meta equilibria. The three strategy profiles can be calculated by exchanging action preferences between different agents, which is called negotiation process. The novel MARL algorithm is called negotiation-based Q-learning.

The full framework of negotiation-based Q-learning algorithm is as follows.

In each state, NegoQ agents may choose a PNE, an EDSP, a non-strict EDSP, or a meta equilibrium. This makes the theoretical

---

Algorithm 1.3: Negotiation for pure strategy NE set

     Input: A normal-form game $\langle N, \{A_i\}_{i=1}^{n}, \{U_i\}_{i=1}^{n} \rangle$

     /* But agent i only knows N, $\{A_i\}_{i=1}^{n}$, and U_i */

1:    The set of pure strategy Nash equilibria $J_{NE} \leftarrow \emptyset$

2:    Compute the maximal utility set $MaxSet_i$

3:    **for each** $\vec{a}_{-i} \in A_{-i}$ **do**

4:       $a_i \leftarrow$ arg $max_{a'_i} U_i(a'_i, \vec{a}_{-i})$

        /* There may be more than one such actions */

5:       $J_{NE} \leftarrow J_{NE} \bigcup \{(a_i, \vec{a}_{-i})\}$

        /* QUESTIONING THREAD: */

6:    **for each** *joint action* $\vec{a} \in J_{NE}$ **do**

7:       Ask all the other agents whether $\vec{a}$ is also in their J_{NE} sets

8:       **if** *one of the answers are 'no'* **then**

9:          $J_{NE} \leftarrow J_{NE} \backslash \{\vec{a}\}$

10:      Tell the other agents to remove $\vec{a}$ from their J_{NE} sets

        /* ANSWERING THREAD: */

11:    **for each** *joint action* $\vec{a}'$ *received from the other agents* **do**

12:      **if** $\vec{a}'$ *is in* $MaxSet_i$ **then**

13:         Send answer 'yes' back to the agent

14:      **else**

15:         Send answer 'no' back to the agent

---

convergence proof of NegoQ difficult. However, Hu *et al.* (2015c) proved the former three strategy profiles belong to the set of symmetric meta equilibria. NegoQ provides a simpler and faster way to learn equilibrium policies without having the knowledge of the other agents' value functions.

1.4 Accelerating MARL by Equilibrium Transfer

Although equilibrium-based MARL can achieve the MARL goal efficiently most of the time, the computation of equilibrium is quite expensive which restrict the scalability heavily. For instance, computing Nash equilibria is polynomial parity arguments on directed graphs ($PPAD$)-hard. When solving programs in correlated equilibrium computation, the number of variables corresponding to the probabilities of taking joint actions increases exponentially with the number of agents. What's more, equilibrium-based MARL requires

---

Algorithm 1.4: Negotiation for non-strict EDSPs set

> **Input:** A normal-form game $\langle N, \{A_i\}_{i=1}^{n}, \{U_i\}_{i=1}^{n}\rangle$, the set
> of pure strategy Nash equilibria J_{NE}
> /* But agent i only knows N, $\{A_i\}_{i=1}^{n}$, and U_i */

1:    The set of non-strict EDSPs $J_{NS} \leftarrow \emptyset$
2:    $X \leftarrow A \backslash J_{NE}$
     /* A is the joint action space */
3:    **for each** *Nash equilibrium* $\vec{e} \in J_{NE}$ **do**
4:      **for each** *joint action* $\vec{a} \in X$ **do**
5:        **if** $U_i(\vec{a}) \geq U_i(\vec{e})$ **then**
6:          $X \leftarrow X \backslash \{\vec{a}\}$
7:          $J_{NS} \leftarrow J_{NS} \cup \{\vec{a}\}$
     /* QUESTIONING THREAD: */
8:    **for each** *joint action* $\vec{a} \in J_{NS}$ **do**
9:      Ask all the other agents whether $\vec{a}$ is also in their J_{NS} sets
10:    **if** *one of the answers are 'no'* **then**
11:      $J_{NS} \leftarrow J_{NS} \backslash \{\vec{a}\}$
     /* ANSWERING THREAD: */
12:    **for each** *joint action* $\vec{a}'$ *received from the other agents* **do**
13:      **if** $\vec{a}'$ *is in* J_{NS} **then**
14:        Send answer 'yes' back to the agent
15:      **else**
16:        Send answer 'no' back to the agent

---

to compute equilibria for a large number of one-shot games for states that should be visited many times until convergence is reached. To reduce the computation load, equilibrium transfer is used in MARL (Hu *et al.*, 2015a).

Equilibrium transfer introduced transfer loss and transfer condition to measure the similarity of two games. During the learning process of equilibrium-based MARL, the one-shot games corresponding to each state's successive visits often have the same or similar equilibria. We can take advantage of this property and reuse the computed equilibria previously when each agent has a small incentive to deviate.

Definition 1.4. An n-agent ($n \geq 2$) one-shot game G is a tuple $\langle N, \{A_i\}_{i=1}^{n}, \{U_i\}_{i=1}^{n}\rangle$, where N is the set of agent and A_i is the action space of agent i.

Algorithm 1.5: Negotiation-based Q-learning algorithm

> **Input:** The agent set N, state space S, and joint action
> space A of a Markov game, learning rate α,
> discount rate γ, exploration factor ϵ

1:    Initialization. $\forall s \in S$ and $\forall \vec{a} \in A, Q_i(s, \vec{a}) \leftarrow 0$
2:    **for each** *episode* **do**
3:        Initialize state s
4:        Negotiate with the other agents according to $Q_i(s)$
5:        $\vec{a} \leftarrow$ the selected equilibrium (with ϵ-greedy)
6:        **repeat**
7:            Receive the experience $(s, \vec{a}, r_i, s')$
8:            Negotiate with the other agents according to $Q_i(s')$
9:            $\vec{a}' \leftarrow$ the selected equilibrium (with ϵ-greedy)
10:           $Q_i(s, \vec{a}) \leftarrow (1 - \alpha)Q_i(s, \vec{a}) + \alpha(r_i + \gamma Q_i(s, \vec{a}'))$
11:           $s \leftarrow s', \vec{a} \leftarrow \vec{a}'$
12:        **until** s *is a terminal state*

Definition 1.5 (Transfer Loss). For a one-shot game G in state s and one of its equilibria p under an equilibrium solution concept Θ, the loss of transfer-ring p to another one-shot game G' in s, ε^{Θ}, is the largest value among all agents' utility losses for not deviating from p in G', which indicates that p is ε^{Θ}-equilibrium ($\varepsilon^{\Theta} \geq 0$) of G'.

To better illustrate transfer loss, we give an instance according to Nash equilibrium. Let $G = \langle N, \{A_i\}_{i=1}^n, \{U_i^G\}_{i=1}^n \rangle$ and $G' = \langle N, \{A_i\}_{i=1}^n, \{U_i^{G'}\}_{i=1}^n \rangle$ be two different one-shot games corresponding to a visit to p^* be a Nash equilibrium of G. According to the definition of NE, the loss of transferring p^* from G to G' is

$$\varepsilon^{NE} = \max_{i \in N} \max_{a_i \in A_i} (U_i^{G'}(a_i, p^*_{-i}) - U_i^{G'}(p^*)) \qquad (1.6)$$

According to the definition of ε^{NE}, for any agent $i \in N$, it holds that

$$U_i^{G'}(p^*) + \varepsilon^{NE} \geq U_i^{G'}(p^*) + \max_{a_i \in A_i} (U_i^{G'}(a_i, p^*_{-i}) - U_i^{G'}(p^*))$$

$$= U_i^{G'}(p^*) + \max_{a_i \in A_i} U_i^{G'}(a_i, p^*_{-i}) - U_i^{G'}(p^*)$$

$$= \max_{a_i \in A_i} U_i^{G'}(a_i, p^*_{-i})$$

When the transfer loss of p is zero, it is an exact equilibrium of G'. However, there are only a few cases where two different one-shot games have the same equilibrium. Instead of requiring zero transfer loss, it is allowed to be a value between zero and a positive threshold τ. By introducing a threshold value, the condition of equilibrium transfer is defined as follows.

Definition 1.6 (Transfer Condition). For two one-shot games G and G', one equilibrium p of G under an equilibrium solution concept Θ, and a real positive value τ, the condition of transferring p from G to G' is that the transfer loss ε^Θ of p is smaller than τ, in which case p^* is at least a τ-approximate equilibrium of G'.

Algorithm 1.6: Equilibrium transfer-based MARL

Input: Learning rate α, discount rate γ, exploration
        factor ϵ,and threshold of transfer loss τ
1: Initialization. $\forall s \in S, \forall i \in N, \forall \vec{a}, Q_i(s, \vec{a}) \leftarrow 0$
2: **for each** *episode* **do**
3:     Initialize state s
4:     **repeat**
5:         $G_c \leftarrow$ the current one-shot game occurring in s
6:         $p^* \leftarrow$ the equilibrium previously computed in s
7:         **if** s *has been visited* **then**
8:           $\varepsilon^\Theta \leftarrow$ the agents' maximum utility loss for transferring p^* to G_c
9:         **else**
10:        $\varepsilon^\Theta \leftarrow +\infty$
11:        **if** $\varepsilon^\Theta > \tau$ **then**
12:          Compute the equilibrium p^* for G_c
13:        **else**
14:          p^* is directly used in G_c
15:        $\vec{a} \leftarrow$ the joint action sampled from p^* (suing ϵ-greedy)
16:        Receive experience $(s, \vec{a}, r_i, s')$ for each $i \in N$
17:        $p' \leftarrow$ the equilibrium stored for the next state s'
18:        **for each** *agent* $i \in N$ **do**
19:          $V_i(s') \leftarrow$ the expected value of p' in s'
20:          $Q_i(s, \vec{a}) \leftarrow (1 - \alpha)Q_i(s, \vec{a}) + \alpha(r + \gamma V_i(s'))$
21:        $s \leftarrow s'$
22:     **until** s *is a terminal state*

By introducing transfer loss and transfer condition into equilibrium-based MARL, the algorithm of equilibrium transfer-based MARL is proposed.

Hu *et al.* (2015a) also has proved that although we use ϵ-equilibrium for calculation, the algorithm can converge perfectly.

1.5 MARL Using Knowledge Transfer

Another effective way to improve scalability of MARL is by using knowledge transfer. Knowledge transfer assumes agents have already learnt some single-agent knowledge (e.g., local value function) before the multiagent learning process. This is inspired by the fact that a lot of MAS only have sparse interactions between agents as we can see in Fig 1.1.

To deal with the sparse interactive MAS mentioned above, Hu *et al.* (2015b) proposed three kinds of knowledge transfer mechanisms:

(1) Value function transfer (VFT).
(2) Selective value function transfer (SVFT).
(3) Transfer-based game abstraction.

1.5.1 *Value Function Transfer*

Among all knowledge transfer mechanisms, VFT is the simplest one. This method transfers agents' local value function to the learning algorithm directly. Since the interaction between agents are sparse, MAS is approximately treated as a set of multiple independent MDPs with one MDP corresponding to one agent. When interaction between agents occurs, agents act as in the single-agent case, then the MARL algorithm is able to learn a coordinating policy for them. Details of the algorithm is shown in Algorithm 1.7.

1.5.2 *Selected Value Function Transfer*

Direct VFT mechanism uses the agents' local value function to initialize the values of joint state–action pairs. This initialization

---

Algorithm 1.7: Value function transfer (VFT)

Input: Learning rate α, discount factor γ, exploration factorϵ, local value function q_i for each agent i

1:   Initialize the state–action value function Q as follows
3:   **for each** *agent* $i \in N$ **do**
3:     **for each** *state* $s \in S$ **do**
4:       **for each** *joint action* $\vec{a} \in A$ **do**
5:         $s_i \leftarrow$ agent i's local state in s
6:         $a_i \leftarrow$ agent i's action in $\vec{a}$
7:         $Q_i(s, \vec{a}) \leftarrow q_i(s_i, a_i)$
8:   The following is the same as general framework of game theory-based MARL

---

helps agents make better choices in most states, but there are still potential risks. For example, when two agents, at the different sides of a narrow alley, are about to reach the other sides, they will make a collision. As a result, neither of the two agents can achieve their goal and they are faced with a huge penalty.

SVFT is proposed to avoid the "unchecked" initialization in direct VFT. For each agent i, a local environmental dynamics is perceived. Whether the knowledge is transferred or not depends on the perception of the agent. It contains two main steps: evaluating the changes of local environmental dynamics and VFT.

1.5.2.1 *Evaluation of local environmental dynamics*

Given a Markov game $M = \langle N, S, \{A_i\}_{i=1}^n, \{R_i\}_{i=1}^n, T \rangle$, we can construct each agent's empirical local environment model by conducting Monte Carlo trials with a random policy. For each agent i, let $M_i = \langle S_i, A_i, R_i^l, T_i^l \rangle$ be its empirical local environment model constructed in the Markov Game M and let $\widehat{M}_i = \langle S_i, A_i, \widehat{R}_i^l, \widehat{T}_i^l \rangle$ be the MDP model of its previous single-agent task. The actual problem is to evaluate how the two MDPs are similar in each state in the local state space S_i.

Definition 1.7. Let $M_i = \langle S_i, A_i, R_i^l, T_i^l \rangle$ be an MDP of agent i. For any two states $s_i, s_i' \in S_i$, the state distance $d_{M_i}(s_i')$ between s_i and

---

Algorithm 1.8: Computing the similarities between two MDPs

Input: Tow MDP models $M_i = \langle S_i, A_i, R_i^l, T_i^l \rangle$ and $\widehat{M}_i = \langle S_i, A_i, \widehat{R}_i^l, \widehat{T}_i^l \rangle$

Output: The set of similarity values $\mathcal{D}$

1: **for each** *state* $s_i \in S_i$ **do**
2:     $\mathcal{D}(s_i) \leftarrow 0$
3:     **for each** *state* $s_i' \in S_i$ **do**
4:         Compute the state distance $d_{M_i}(s_i, si')$ and $d_{\widehat{M}_i}(s_i, s_i')$ according to above function
5:         $\mathcal{D}(s_i) \leftarrow \mathcal{D}(s_i) + (d_{M_i}(s_i, s_i') - d_{\widehat{M}_i}(s_i, s_i'))^2$
6:     $\mathcal{D}(s_i) \leftarrow \sqrt{\mathcal{D}(s_i)}$

---

s_i' is defined as

$$d_{M_i}(s_i, s_i') = \max_{a_i \in A_i} \{ |R_i^l(s_i, a_i) - R_i^l(s_i', a_i)|$$
$$+ \gamma \mathcal{T}_{d_{M_i}}^K (T_i^l(s_i, a_i), T_i^l(s_i', a_i)) \} \tag{1.7}$$

where γ is the discount factor, and $\mathcal{T}_{d_{M_i}}^K (T_i^l(s_i, a_i), T_i^l(s_i', a_i))$ is the Kantorovich distance between the probabilistic distributions $T_i^l(s_i, a_i)$ and $T_i^l(s_i', a_i)$. Based on the state distance defined above, the MDP distance concept is defined as follows.

Definition 1.8. Given two MDPs $M_i = \langle S_i, A_i, R_i^l, T_i^l \rangle$ and $\widehat{M}_i = \langle S_i, A_i, \widehat{R}_i^l, \widehat{T}_i^l \rangle$, for any state $s_i \in S_i$, the similarity between M_i and $\widehat{M}_i$ in s_i is defined as

$$\mathcal{D}_{M_i, \widehat{M}_i}(s_i) = \sqrt{\sum_{s_i' \in S_i} (d_{M_i}(s_i, s_i') - d_{\widehat{M}_i}(s_i, s_i'))^2} \tag{1.8}$$

1.5.2.2 *The SVFT algorithm*

There are three steps in SVFT. Firstly, for each agent i and empirical MDP model, M_i by Monte Carlo sampling in the Markov game is constructed. Secondly, for each agent i, the similarities between M_i and the MDP in its previous single agent task $\widehat{M}_i$ are computed. Lastly, the local value function of each agent is transferred only

---

Algorithm 1.9: Selective value function transfer

Input: Learning rate α, discount factor γ, exploration factor ϵ, local value function q_i and local MDP model $\widehat{M}_i$ for each agent i, a threshold value τ, and integer L for Monte Carlo sampling

1: **for** *episode* $= 1, \ldots, L$ **do**
2:     Perform Monte Carlo sampling with a random policy, recording the rewards and state transitions
3: **for each** *agent* $i \in N$ **do**
4:     $M_i \leftarrow$ the empirical MDP model of agent i
5:     $\mathcal{D}_i \leftarrow$ the set of similarities between M_i and $\widehat{M}_i$
6: Transfer the local value functions as follows
7: **for each** *agent* $i \in N$ **do**
8:     **for each** *state* $s \in S$ **do**
9:         **for each** *joint action* $\vec{a} \in A$ **do**
10:             $s_i \leftarrow$ agent i's local state in s
11:             $a_i \leftarrow$ agent i's action in $\vec{a}$
12:             **if** $\mathcal{D}_i(s_i) \leq \tau$ **then**
13:                 $Q_i(s, \vec{a}) \leftarrow q_i(s_i, a_i)$
14: The following is the same as general framework of game theory-based MARL

---

in the local states where the similarity is small. The corresponding algorithm is shown in Algorithm 1.9.

1.5.3 *Model Transfer-based Game Abstraction*

VFT and SVFT can modify the initialization of value functions in game theory-based MARL. However, the learning process is also computationally expensive for we should compute an equilibrium in each visited state. In the case of sparse interactive MAS, there is no use calculating equilibrium all the time. Sometimes there is no game between the agents or the game does not involve all the agents. To this end, the idea of game abstraction is proposed.

Model Transfer-based Game Abstraction (MTGA) abstracts the one-shot game in each state into a smaller one by removing the agents which do not join in the game. Thus, the computation of equilibrium is simplified. If the local environmental dynamics of agent i in state s are very similar to those in its previous single-agent task, then it can

---

Algorithm 1.10: Model transfer-based game abstraction (MTGA)

    Input: Learning rate α, discount factor γ, exploration factor ϵ,
                local MDP model $\widehat{M_i}$ for each agent i, a threshold value
                τ, an integer L for Monte Carlo sampling

1: Initialize the joint state–action value function, $\forall i \in N, Q_i \leftarrow \phi$
2: Initialize the local state–action value function, $\forall i \in N, \forall s_i \in S_i$,
    $\forall a_i \in A_i, q_i(s_i, a_i) \leftarrow 0$
3: **for each** *episode* **do**
4:     Initialize state s
5:     **repeat**
6:        $X \leftarrow \{i \in N \mid \mathcal{D}_i(s_i) > \tau\}$
7:        $s_g \leftarrow$ the joint state of the agents in X
8:        **for each** *agent* $i \in X$ **do**
9:           **if** s_g *is not in the state space of* Q_i **then**
10:             **for each** *joint action* a_g of the agents in X **do**
11:                Extend Q_i to include (s_g, a_g)
12:                $Q_i(s_g, a_g) \leftarrow 0$
13:        $G_X(s) \leftarrow \langle X, \{A_i\}_{i \in X}, \{Q_i(s)\}_{x \in X} \rangle$
14:        **for each** *agent* $i \notin X$ **do**
15:           $a_i \leftarrow \arg\max_{a'_i} q_i(s_i, a'_i)$
16:        **for each** *agent* $i \in X$ **do**
17:           $a_i \leftarrow$ the action sampled by the equilibrium of
            $G_X(s)$ for agent i
18:        $\vec{a} \leftarrow (a_1, \ldots, a_n)$
19:        Receive the experience $(s, \vec{a}, r_i, s')$ for each agent i
20:        **for each** *agent* $i \notin X$ **do**
21:           $q_i(s_i, a_i \leftarrow (1 - \alpha)q_i(s_i, a_i) + \alpha(r_i + \gamma\Phi_i(s^{\cdot}))$
22:        **for each** *agent* $i \in X$ **do**
23:           $a_g \leftarrow$ the joint action taken by the agents in X
24:           $Q_i(s_i, a_i \leftarrow (1 - \alpha)Q_i(s_i, a_i) + \alpha(r_i + \gamma\Phi_i(s^{\cdot}))$
25:        $s \leftarrow s'$
26:     **until** s *is a terminal state*

---

be removed out of the game. Based on the concept of MDP similarity, the game abstraction is defined as follows.

Definition 1.9. Given a Markov game $M = \langle N, S, \{A_i\}_{i=1}^n, \{R_i\}_{i=1}^n, T \rangle$. Let $G(s) = \langle N, \{A_i\}_{i=1}^n, \{Q_s\}_{i=1}^n \rangle$ be the one-shot game in any state $s \in S$. Define a subset of N by $X = \{i \in N | \mathcal{D}_i(s_i) > \tau\}$, where s_i is the local state of agent i in state s, $\mathcal{D}_i(s_i)$ is the

MDP similarity defined above, and τ is a threshold value. The abstracted game of $G(s)$ derived from X is defined as $G_X(s) = \langle X, \{A_i\}_{i \in X}, \{Q_i(s)\}_{i \in X} \rangle$.

The set X contains all agents that are considered to be related in the state s. If $|X| < |N|$, then the abstracted game $G_X(s)$ has a smaller size than $G(s)$.

VFT conducts knowledge transfer in all states, but SVFT only transfers knowledge in states where the agents' local environmental dynamics are similar to those in the previous single-agent tasks. Compared with VFT and SVFT, MTGA does not need to compute an equilibrium in most of the states and does not require observability of all joint states and actions. Since VFT and SVFT only modify the process of value function initialization, their convergence properties totally depend on the corresponding learning algorithms. For the mechanism of MTGA, there is no formal convergence guarantees currently. It is an interesting future direction to theoretically analyze the convergence property of MTGA.

Chapter 2

Centralization or Decentralization? A Compromising Solution Toward Coordination in Multiagent Systems*

Chao Yu[†], Hongtao Lv[‡], Hongwei Ge, Liang Sun,
Jun Meng and Bingcai Chen

*School of Computer Science and Technology,
Dalian University of Technology, Dalian 116024, China*

[†] *cy496@dlut.edu.cn*
[‡] *lvhongtao@mail.dlut.edu.cn*

Coordination of agent behaviors is crucial in multiagent systems (MASs). Social norm has been considered to be an effective tool to achieve coordination in MASs by placing social constraints on agent actions. The two distinct approaches for achieving social norms, i.e., the centralized top-down prescriptive approach and the decentralized bottom-up emerging approach, have their own inevitable advantages and disadvantages. How to make the best of these two approaches and circumvent their drawbacks is a challenging problem in norm research. In this chapter, we propose a hierarchical learning framework that creates a balance between centralized control and distributed interactions during norm emergence by integrating hierarchical supervision into distributed learning interactions. Experiments are carried out to explore the effectiveness of the proposed framework in

---

*This chapter is a significantly extended version of previous publication: C. Yu, H. Lv, F. Ren, H. Bao and J. Hao, Hierarchical Learning for Emergence of Social Norms in Networked Multiagent Systems, *In Proc. AI*, pp. 630–643, 2015.

various situations, and results verify that this compromising solution can be an effective mechanism for achieving coordination in MASs.

2.1 Introduction

One of the most critical problems in the coordinated control of large-scale distributed multiagent systems (MASs) is to design efficient strategies that enable all the agents to reach an agreement in areas of common interest (Wu *et al.*, 2012). The concept of social norms (Shoham and Tennenholtz, 1997), originally used in the field of sociology to study human social behavior, is of great interest to MAS researchers as it can be used to help increase the predictability of agent behavior, and thus facilitate coordination among agents in the whole system. There have been numerous theoretical investigations in the MAS literature of social norms under different assumptions about agent interaction protocols, societal topologies, and observation capabilities (Hao *et al.*, 2015; Mukherjee *et al.* 2008; Sen and Airiau, 2007; Villatoro *et al.*, 2011; Yu *et al.*, 2014, 2015a, 2016a). In empirical applications, social norms have been used as an efficient mechanism to coordinate agent behaviors in large-scale distributed systems such as electronic institutions (Criado *et al.*, 2011b), norm-supported computational societies (Artikis *et al.*, 2009), and normative *ad hoc* networks (Artikis *et al.*, 2004).

It has been well recognized that two distinct approaches are available for the establishment of a social norm in MASs (Savarimuthu and Cranefield, 2011). The first one is the centralized prescriptive approach that is formulated by an omnipresent authority which specifies and enforces how the agents should behave according to the administrative incentives. The second one is the decentralized bottom-up approach that enables a norm to evolve and emerge on its own without relying on any centralized authority. The former is often based on the offline design, where every agent has the norms "hard-wired" at the beginning, while the latter is usually based on the online emergent design, where agents decide the most suitable conventions through their local interactions (Delgado, 2002).

These two different approaches for achieving social norms have their own advantages and disadvantages. Although the centralized approach is capable of achieving the best system performance, this optimal performance is usually at the expense of high administrative cost (e.g., communication cost). Moreover, as the environments where agents are located become dynamic and open, and the system may involve a large number of distributed agents, it is usually expensive or sometimes infeasible to have a centralized enforcer to formulate and specify social norms in a prescriptive manner. So, for many large-scale systems, it is more desirable to enable social norms to evolve on their own automatically. However, this decentralized approach has its own limitation in terms of low efficiency due to distributed interactions among agents. Therefore, it is necessary to create a balance between these two kinds of approaches so as to achieve the maximal system performance while bounding the administrative cost at a proper level.

To this end, we propose a hierarchical learning framework that features the combination of centralized governing and decentralized interactions. In the framework, agents are separated into different clusters. Each agent in a cluster interacts with others locally using reinforcement learning (RL) methods, and reports its learning information to an upper-layer supervisor in the cluster. After synthesizing all the interaction information in the cluster, the supervisor then generates a supervision policy through interacting with its neighboring supervisors, and passes down this policy to subordinate agents in order to adapt their learning strategies heuristically. The highlight of the proposed framework is that through hierarchical supervision between subordinate agents and supervisors, a compromising solution can be made to elegantly balance distributed interactions and centralized control for norm emergence. Experiments show that this kind of compromising solution can be an effective mechanism for emergence of social norms.

The remainder of the chapter is organized as follows. Section 2.2 introduces social norms and RL. Section 2.3 introduces the proposed learning framework. Section 2.4 presents experimental studies.

Section 2.5 discusses related work. Finally, Section 2.6 concludes this chapter.

2.2 Social Norms and RL

As most previous studies do, this chapter also uses learning "rules of the road" (Sen and Airiau, 2007; Young, 1996) as a metaphor to study emergence of norms, which can be viewed as a coordination game (CG) in Table 2.1. In CG, the agents are positively rewarded when both of them have chosen the same action, and penalized otherwise. The abstraction of coordination given by the CG covers a number of practical scenarios, such as distributed robots coordinating on which object to work on together, and wireless nodes coordinating on which channel to transmit messages (Mihaylov *et al.*, 2014). The problem to deal with the CG is that there is nothing in the structure of the game itself that allows players (even purely rational players) to infer what they ought to do (Young, 1996). Therefore, social norms can be used to guide agent behaviors towards specific ones when moral or rational reasoning does not provide a clear guidance of how to behave. To study the impact of norm space on norm emergence, the CG can be extended to a general form by considering N_a actions in the norm space, as shown in Table 2.2.

RL algorithms (Sutton and Barto, 1998b) have been widely used for agent interactions in previous research on norm emergence. We focus on the Q-learning algorithm in this chapter, in which an agent makes a decision through the estimation of a set of Q-values. Its one-step updating rule is given by Eq. (2.1) (Watkins and Dayan, 1992),

$$Q(s,a) = Q(s,a) + \alpha_i [r(s,a) + \gamma \max_{a'} Q(s',a') - Q(s,a)] \quad (2.1)$$

Table 2.1. The CG.

	L	R
L	1, 1	−1, −1
R	−1, −1	1, 1

Table 2.2. The extended CG.

	a_1	a_2	$\cdots$	a_{N_a}
a_1	1,1	$-1,-1$	$\cdots$	$-1,-1$
a_2	$-1,-1$	1,1	$\cdots$	$-1,-1$
$\vdots$	$\vdots$	$\vdots$	$\ddots$	$\vdots$
a_{N_a}	$-1,-1$	$-1,-1$	$\cdots$	1,1

where $\alpha_i \in (0,1]$ is a learning rate of agent i, $\gamma \in [0,1)$ is a discount factor, $r(s,a)$ and $Q(s,a)$ are the immediate and expected reward of choosing action a in state s at time step t, respectively, and $Q(s',a')$ is the expected discounted reward of choosing action a' in state s' at time step $t+1$.

Q-values of each state–action pair are stored in a table for a discrete state–action space. At each time step, agent i chooses the best-response action with the highest Q-value based on the corresponding Q-value table with a probability of $1-\epsilon$ (i.e., exploitation), or chooses other actions randomly with a trial-and-error probability of ϵ (i.e., exploration).

2.3 The Proposed Hierarchical Learning Framework

2.3.1 *The Principle of the Learning Framework*

In the proposed framework, agent interactions are purely local and are constrained by the network structures. The agent network is separated into a series of clusters C_x, $x \in (1,2,\ldots,X)$, where X is the number of clusters, and x denotes a supervising agent (supervisor) for each cluster. The supervisor can be any one of the subordinate agents in the cluster or another dedicated agent. But to simplify calculation and illustration, we consider that the supervisor of a cluster is situated in the geometric center of this cluster. The supervisors are also connected with each other based on the lower network structure, and they can also interact with their neighbors. An example of the hierarchical network structure is given by Fig. 2.1.

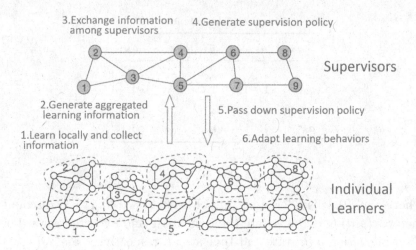

Fig. 2.1.   Illustration of the hierarchical learning framework.

Based on Q-learning, the interaction protocol of the proposed framework in Algorithm 2.1 is briefly described as follows:

(1) At each time step t, agent i chooses an action a_i with the highest Q-value or randomly chooses an action with an exploration probability ϵ_i^t. Agent i then interacts with a random neighbor j and receives a payoff of r_i;

(2) The learning experience in terms of action–reward pair (a_i, r_i) is reported to agent i's supervisor x, and supervisor x combines all actions of its subordinate agents into a supervision policy a_x;

(3) Supervisor x then interacts with one of its neighboring supervisors and updates a_x based on the performance difference between the action of supervisor x and that of its neighbor; The updated a_x', which is considered to be the most successful action in the neighborhood, is then issued to its subordinate agents;

(4) Based on the updated supervision policy, agent i adjusts its learning behaviors in terms of learning rate α_i^t and/or the exploration rate ϵ_i^t accordingly;

(5) Finally, agent i updates its Q-value using the new learning rate α_i^{t+1}.

Algorithm 2.1 Interaction protocol of the hierarchical learning model

---

for *each step t* $(t = 1, \ldots, T)$ **do**

    for *each agent i* $(i = 1, \ldots, n)$ **do**

        Chooses an action a_i based on its Q-values with exploration ϵ_i^t; interacts with a neighbor j and gets a reward r_i; reports a_i and r_i to its supervisor;

    end

    for *each supervisor x* $(x = 1, \ldots, X)$ **do**

        Combines the actions of its all subordinate agents into a supervision policy a_x; interacts with one of its neighboring supervisor y and converts a_x to a_y with a probability $p_{x \to y}$; passes down updated supervision policy a'_x to its subordinate agents;

    end

    for *each agent i* $(i = 1, \ldots, n)$ **do**

        Updates α_i^t / ϵ_i based on the a'_x from its supervisor; updates Q-values using new learning rate (i.e., α_i^{t+1});

    end

end

The main feature of the hierarchical learning model is that there is hierarchical supervision governing over the distributed interactions among agents. As supervisors have a wider view than subordinate agents, they can direct their subordinate agents by learning from other supervisors. Through the feedback from supervisors, subordinate agents can understand whether their actions comply with the norms in the society, and thereby adjust their learning behaviors accordingly. By integrating supervisors' central control into agents' local learning processes, norms can be established more effectively and efficiently in the hierarchical learning framework, compared with the traditional learning framework based on pure distributed interactions.

2.3.2 *Generation of Supervision Policies*

Each focal agent i reports its action a_i and payoff r_i to its supervisor x. The supervisor aggregates all the information from its subordinates into two values, i.e., F_x and R_x. Value $F_x(a)$ indicates the overall acceptance (i.e., frequency) of action a in cluster C_x, and value $R_x(a)$ indicates the overall reward of action a in the cluster. $F_x(a)$ can be calculated by Eq. (2.2),

$$F_x(a) = \sum_{i \in C_x} \delta(a, a_i) \tag{2.2}$$

where $\delta(a, a_i)$ is the Kronecker delta function, which equals to 1 if $a = a_i$, and 0 otherwise.

$R_x(a)$ can be calculated by Eq. (2.3),

$$R_x(a) = \frac{1}{F_x(a)} \sum_{i \in C_x, a_i = a} r_i \tag{2.3}$$

especially, $R_x(a)$ is set to 0 if $F_x(a) = 0$.

Each supervisor x then combines the actions of each cluster into a supervision policy a_x using democratic voting mechanism, which means that a_x is simply the action most accepted by the cluster (randomly choose an action if there exists a tie). The supervision policy a_x can be given by $a_x = \text{argmax}_a F_x(a)$.

After combining actions of the cluster, each supervisor generates a new supervision policy a'_x, which indicates the social norm (i.e., the majority action that has been adopted) accepted by the cluster. In order to generate a better supervision policy, each supervisor then interacts with a randomly chosen neighboring supervisor and learns from the neighbor by comparing the performance of their actions. The motivation of this comparison comes from evolutionary game theory (EGT) (Szabó and Fáth, 2006), which provides a powerful methodology to model how strategies evolve over time based on their performance. In this theory, strategies compete with each other and evolve while agents learn (mainly through imitation) from others with higher fitness. One of the widely used imitation rules is the proportional imitation, which can be given by Eq. (2.4),

$$p_{x \to y} = \frac{1}{1 + e^{-\beta(u_y - u_x)}} \tag{2.4}$$

where $p_{x \to y}$ is a probability for supervisor x to imitate the action of neighbor y, $u_x = R_x(a_x)$ means the fitness (average payoff of the supervision policy) of supervisor x, $u_y = R_y(a_y)$ means the fitness of neighboring supervisor y, and $\beta > 0$ is a parameter to control selection bias.

In Eq. (2.4), probability $p_{x \to y}$ becomes larger when the gap between u_x and u_y becomes smaller. Every supervisor x then converts a_x into a_y as the final supervision policy with probability $p_{x \to y}$, or remains a_x with probability $1 - p_{x \to y}$. In this way, supervisors imitate better actions taken by their neighbors to improve the performance of their clusters.

2.3.3 *Adaption of Local Learning Behaviors*

Based on the principle of EGT, a supervision policy represented as the new action a'_x is generated. The new action a'_x indicates the most successful action in the neighborhood and therefore should be integrated into the learning process in order to entrench its influence. By comparing its action at time step t (i.e., a_i^t) with the guiding policy a'_x, agent i can evaluate whether it is performing well or not so that its learning behavior can be dynamically adapted to fit the supervision policy. Depending on the consistency between the agent's action and the supervision policy, the agent's learning process can be adapted according to the following three mechanisms:

- **Hierarchial Learning-α (*HL-α*):** In RL, the learning performance heavily depends on the learning rate parameter, which is difficult to tune. This mechanism adapts the learning rate α in the learning process. When agent i has chosen the same action with the supervision policy (i.e., $a_i^t = a'_x$), it decreases its learning rate to maintain its current state, otherwise it increases its learning rate to learn faster from its interaction experience. Formally, learning rate α_i^t can be adjusted according to

$$\alpha_i^{t+1} = \begin{cases} (1 - \lambda)\alpha_i^t & \text{if } a_i^t = a'_x \\ (1 - \alpha_i^t)\alpha_i^t + \lambda & \text{otherwise} \end{cases} \qquad (2.5)$$

  where $\lambda \in [0, 1]$ is a parameter to control the adaption rate.
- **Hierarchial Learning-ϵ (*HL-ϵ*):** Exploration–exploitation trade-off has a crucial impact on the learning process. Therefore, this mechanism adapts the exploration rate ϵ in the learning process. The

motivation of this mechanism is that an agent needs to explore more of the environment when it is performing poorly and explore less otherwise. Similarly, the exploration rate ϵ_t^t can be adjusted according to

$$\epsilon_i^{t+1} = \begin{cases} (1 - \lambda)\epsilon_i^t & \text{if } a_i^t = a_x' \\ \min\{(1 - \lambda)\epsilon_i^t + \lambda, \overline{\epsilon_i}\} & \text{otherwise} \end{cases} \quad (2.6)$$

in which $\overline{\epsilon_i}$ is a variable to confine the exploration rate to a small value in order to indicate a small probability of exploration in RL.

- **Hierarchical Learning-$\alpha \cdot \epsilon$ (HL-$\alpha \cdot \epsilon$):** This mechanism adapts the learning rate and the exploration rate at the same time based on HL-α and HL-ϵ.

Learning rate and exploration rate are two fundamental tuning parameters in RL. Heuristic adaption of these two parameters thus models the adaptive learning behavior of agents. The proposed mechanisms are based on the concept of "winning" and "losing" in the well-known MAL algorithm Win-or-Learn-Fast (WoLF) (Bowling and Veloso, 2002). Although the original meaning of "winning" or "losing" in WoLF and its variants is to indicate whether an agent is doing better or worse than its Nash Equilibrium policy, this heuristic is gracefully introduced into the proposed model to evaluate the agent's performance against the supervision policy. Specifically, an agent is considered to be winning (i.e., performing well) if its action is the same with the supervision policy, and losing (i.e., performing poorly) otherwise. The different situations of "winning" or "losing" thus indicate whether the agent's action is complying with the norm in the society. If an agent is in a losing state (i.e., its action is against the norm in the society), it needs to learn faster or explores more of the environment in order to escape from this adverse situation. On the contrary, it should decrease its learning and/or exploration rate to stay in the winning state.

2.3.4 *Price of Anarchy and Monarchy*

In the hierarchical learning framework, the size of clusters is an important indicator to measure the level of centralization. In order to explicitly quantify the level of centralization, we define two criteria, i.e., the price of anarchy (PoA) and the price of monarchy (PoM). The quantification of these two values provides important criteria for coordination solutions, that is, how to choose a proper cluster size in the hierarchical learning framework.

The PoA, which was first introduced by (Koutsoupias and Papadimitriou, 1999), measures the inefficiency of a selfish equilibrium. This conventional definition was then extended to define the inefficiency of a multiagent learning algorithm (Oh and Smith, 2008). Here, we generalize this definition further to measure the ineffectiveness of a distributed solution against a centralized optimal solution to a coordination problem. More formally, the PoA can be given by

$$\text{PoA} = \frac{\psi_{\text{opt}} - \psi_{\text{dis}}}{\psi_{\text{opt}}}, \quad \text{PoA} \in [0, 1] \tag{2.7}$$

where ψ_{opt} is the optimal performance using a centralized solution and ψ_{dis} is the performance of a distributed solution. The performance means any criterion that evaluates a solution, e.g., convergence level/ratio/speed for an emergence process. It can be seen that a lower PoA indicates a more effective distributed solution to a coordination problem.

Analogous to the PoA, we can define the PoM. Whereas the PoA measures potential quality loss due to distributed solution, the PoM estimates the practical cost of maintaining centralization in a system. To simplify the illustration, we mainly discuss managerial cost in terms of communication cost. Thus, the lower bound of the PoM is found in a fully distributed non-communicating system, and the upper bound of the PoM is found in a fully centralized system that relies on communication to implement a coordination mechanism. Formally, let φ_{dis} and φ_{opt} denote a communication cost function of a distributed solution and a centralized solution, respectively. The PoM is given by

$$\text{PoM} = \frac{\varphi_{\text{dis}}}{\varphi_{\text{opt}}}, \quad \text{PoM} \in [0, 1] \tag{2.8}$$

It is obvious that a distributed solution with lower PoM requires lower communication cost to realize coordination of a system.

2.4   Experiments and Results

In this section, experiments are carried out to demonstrate the performance of the proposed hierarchical learning framework. Unless otherwise specified, the network topology is a 10×10 grid network separated into several 4×4 clusters.[1] In the Q-learning algorithm, we consider each state as the same, which means that there is no state transition. Each agent can choose from four actions as default. Parameters α and ϵ are initially set as 0.1 and 0.01, respectively. Moreover, parameters β and γ are both set as 0.1. The final results are averaged over 10,000 independent runs.

In order to demonstrate the benefits of hierarchical supervision in the proposed learning framework, we compare the performance between the approach based on traditional Individual Learning (IL) framework (Sen and Airiau, 2007) and the three approaches based on our hierarchical learning framework. The IL learning approach is a fully decentralized approach in terms that agents learn randomly with another agent in the population and update their learning strategies individually. Figure 2.2 presents the learning dynamics of the different learning approaches, in which IL-Decaying α and IL-Decaying ϵ represent IL approaches with a decaying learning rate and exploration rate, respectively, and IL-fixed $\alpha \cdot \epsilon$ represents IL approach with a fixed learning rate and exploration rate. The result shows that the average reward of the system increases as time proceeds, which means that social norms can emerge using all these approaches. The convergence level and efficiency, however, differ among these approaches. The hierarchical learning approaches (especially HL-ϵ and HL-$\alpha \cdot \epsilon$) are slower at first but finally reach a higher convergence level than the other approaches. This result indicates that with the supervision policy from supervisors, agents can know more about the society, and the hierarchical mechanism is an effective way to remove subnorms in order to reach a higher level of coordination.

---

[1]The 10×10 grid network is divided into 4×4 clusters, and the remaining two agents on the border are included in a single cluster.

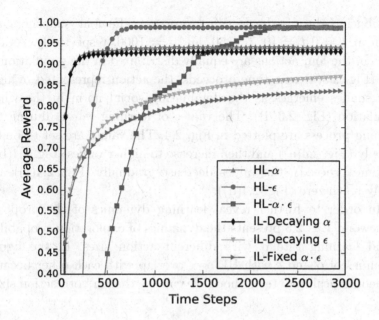

Fig. 2.2. Comparison when learning rate decreases.

Table 2.3. Learning performance with different combination of adapting ϵ and α.

HL-ϵ (%)	HL-α (%)					
	0	0.2	0.4	0.6	0.8	1
0	0.867014	0.910901	0.921715	0.927379	0.929733	0.936224
0.2	0.862955	0.905489	0.923335	0.929961	0.938674	0.945026
0.4	0.919847	0.947869	0.959869	0.964609	0.969008	0.971451
0.6	0.972477	0.975971	0.978973	0.979797	0.980279	0.980726
0.8	0.98434	0.98437	0.984517	0.98454	0.984566	0.984492
1	0.988187	0.988158	0.988041	0.988102	0.98807	0.98819

Table 2.3 gives the detailed performance when agents have different probabilities to choose among HL-α and HL-ϵ. We can see that the simultaneous adaption of α and ϵ can further facilitate norm emergence in the system.

In order to give a vivid illustration of the dynamics under the hierarchical learning framework, Fig. 2.3 plots the snapshot of action distribution during the learning process. The four snapshots are taken at $t = 0, t = 10$, $t = 50$, and $t = 1000$, respectively. As can be seen, the four actions are equally distributed in the population at first (Fig. 2.3(a)). As time proceeds, the action represented by light gray shades emerges as the dominating social norm in the whole population (Fig. 2.3(d)). The values of learning rate α during the learning process are plotted in Fig. 2.4. The values are set to be 0.1 initially (Fig. 2.4(a)) and then increase to higher values (Fig. 2.4(b)). As time proceeds, the values decrease gradually (Fig. 2.4(c)) and finally reach zero (Fig. 2.4(d)).

In order to further reveal learning dynamics of the proposed framework, Fig. 2.5 presents the dynamics of exploration probability ϵ and learning rate α with different action sizes. In the figure, dynamics of α and ϵ with HL-$\alpha \cdot \epsilon$ overlap with each other because of the identical update method. We can see that in both action sizes,

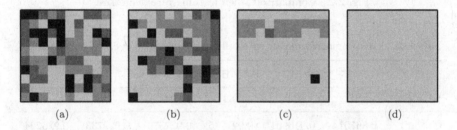

(a)   (b)   (c)   (d)

Fig. 2.3.   The snapshot of action distribution.

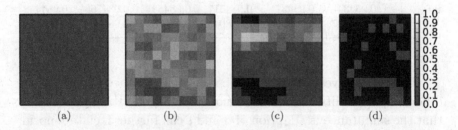

(a)   (b)   (c)   (d)

Fig. 2.4.   The snapshot of values of learning rate α.

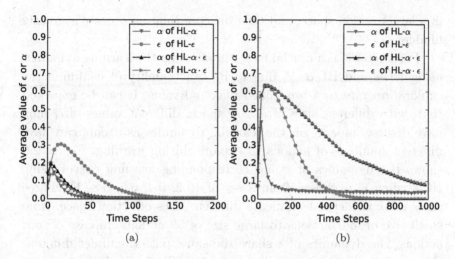

Fig. 2.5. Dynamics of ϵ and α with (a) two actions and (b) four actions.

the values of α and ϵ increase sharply at the beginning, and then drop gradually to near zero. This is because the whole agent system is still in chaos at the beginning of learning as agents are not sure which action is the best choice and thus choose their actions randomly. In this case, it is more likely that the agents are in a "losing" state caused by failed interactions among the agents. In order to get over the "losing" state, agents would increase their learning rate and/or exploration rate to learn faster and/or explore more from the interactions. As the process moves on, each agent's action is more and more consistent with its supervision policy. Thus, ϵ and α decrease accordingly to indicate a "winning" state of the agents. The difference between Figs. 2.5(a) and 2.5(b) indicates that, in a four-action scenario, the values change more drastically at first, and then it takes a longer time for these values to decrease to zero. This is because agents are more likely to choose the same action for achieving a coordination in a smaller size of action space. When the number of actions gets larger, the probability to find the right action as the consensus is greatly reduced. The large number of conflicts among the agents thus cause the agents to be in a "losing" state more often in a larger action space, and thus the coordination process is greatly hindered.

Parameter $\overline{\epsilon_i}$ is a crucial factor in affecting the learning dynamics using HL-ϵ and HL-$\alpha \cdot \epsilon$, due to its functionality of confining the exploration rate to a predefined maximal value. It can be expected that, with different sizes of action space, different values of $\overline{\epsilon_i}$ may have diverse impacts on the learning dynamics as agents can have different numbers of actions to explore during learning. Figure 2.6 shows the dynamics of ϵ and corresponding learning curves using HL-ϵ when $\overline{\epsilon_i}$ is chosen from a set of $\{0.2, 0.4, 0.6, 0.8, 1\}$. Three cases are considered to indicate different sizes of action space, from small size of four actions to large size of 50 actions. In case of four actions, the dynamics of ϵ share the same patterns under different values of $\overline{\epsilon_i}$. The values spike sharply at the beginning process of learning, and then drop gradually to zero. The peaks of ϵ, however, differ from each other, from around 0.18 when $\overline{\epsilon_i} = 0.2$ to around 0.63 when $\overline{\epsilon_i} = 1$. This is because a larger $\overline{\epsilon_i}$ enables the agents to explore more action choices during learning. Higher exploration accordingly causes more failed interactions among the agents and thus the exploration rate ϵ will increase further to indicate a "losing" state of the agent. The corresponding learning curves in terms of average rewards of agents indicate that the coordination process is hindered when using a small value of $\overline{\epsilon_i}$. The dynamics patterns, however, are quite different in cases of 10 and 50 actions. In these two scenarios, the values of ϵ cannot converge to zero when $\overline{\epsilon_i} = 1$ and 0.8 in 10^4 time steps. This is because agents have a large number of alternatives to explore during the learning process, which can cause the agents to be in a state of "losing" constantly. This accordingly increases the values of ϵ until reaching the maximal values of $\overline{\epsilon_i}$. As a result, coordination cannot be achieved among the agents, which can also be observed from the low level of average rewards at the bottom low of Fig. 2.6. Although ϵ can gradually decline to zero when $\overline{\epsilon_i} = 0.6$, 0.4, and 0.2, the learning dynamics in these three cases vary a bit. The norm emergence processes are slower at first when $\overline{\epsilon_i} = 0.6$, but then catch up with those when $\overline{\epsilon_i} = 0.4$ and 0.2, and then keep faster afterwards. The general results revealed

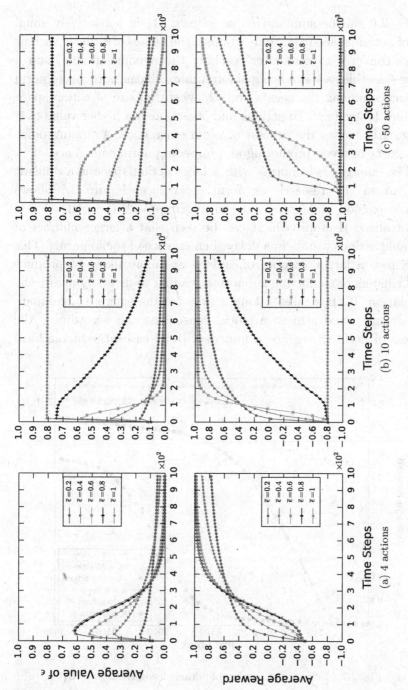

Fig. 2.6. Dynamics of ϵ with different values of $\overline{\epsilon_i}$.

in Fig. 2.6 can be summarized as follows: (1) in a relatively small size of action space (e.g., four actions), the values of ϵ under various $\overline{\epsilon_i}$ can converge to zero after reaching the maximal points, and a larger $\overline{\epsilon_i}$ in this case can bring about a more efficient process of norm emergence among the agents and (2) when the size of action space becomes larger (e.g., 10 actions and 50 actions), a higher value of $\overline{\epsilon_i}$ can greatly hinder the process of norm emergence. A tipping point of $\overline{\epsilon_i}$ exists between promoting and impeding norm emergence.

The emergence of norms with a large action space is a difficult problem in the research of norm emergence. Figure 2.7 shows the comparison of hierarchical learning approach (i.e., HL-α) and decentralized IL approach. It can be seen that a larger number of available actions results in a delayed emergence of social norms. This is because a larger number of actions are more likely to produce local clusters of conflicting subnorms, leading to diversity across the population. It thus takes a longer time for the agents to eliminate this diversity and achieve a global consensus, and accordingly the process of norm emergence is impeded. In all cases, the hierarchical

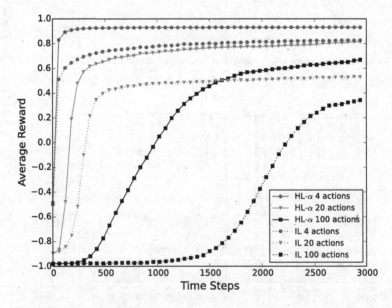

Fig. 2.7.　Influence of number of actions using HL-α and IL.

learning approach performs better than the IL approach in terms of a faster convergence speed and a higher convergence level. This is because the agents can receive support from the wide view of their supervisors and get access to the state of the whole society. With this information, they can understand whether they have complied with the norm in the society and decide to learn/explore more or to keep their actions. This result shows that the proposed hierarchical learning model is indeed effective for achieving norm emergence in a large action space.

The influence of population size on norm emergence is shown in Fig. 2.8. In both approaches of HL-α and IL, the convergence process is hindered as the population is growing larger. This result is because the larger the society, the more difficult to diffuse the effect of local learning to the whole society. This phenomenon can be observed in human societies where small groups can more easily establish social norms than larger groups. The proposed hierarchical learning approach HL-α, however, can greatly facilitate norm emergence in different population sizes. This is because in the hierarchical

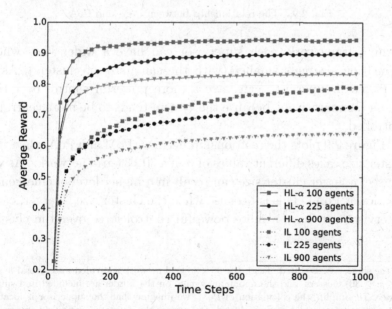

Fig. 2.8. Influence of population size using HL-α and IL.

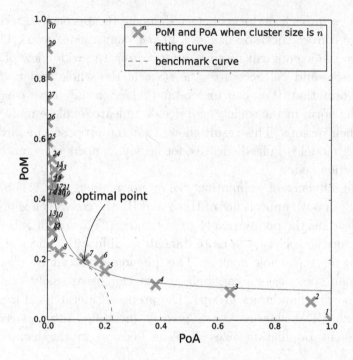

Fig. 2.9.   The relationship between PoM and PoA.

learning framework, each supervisor can have a wider view when population size gets smaller (with the same size of clusters). As a result, each supervisor can have a more powerful control over the cluster comparatively, because the system tends to be more centrally controlled.

Figure 2.9 plots the relationship between PoM and PoA when the cluster size takes different values of n in a 30×30 grid network.[2] As we can see, a larger cluster size can result in a higher level of emergence (i.e., lower PoA). This is because when the cluster size is larger, each supervisor can have a more powerful control force over the cluster

---

[2]In the same way as stated ahead, the 30×30 grid network is divided into several $n \times n$ ($1 \leq n \leq 30$) clusters, and the remaining agents on the border are included in a single cluster. To simplify the calculation of PoM, we imagine that the supervisor is located in the geometrical center of each cluster, and we use the geometric distance between an agent and the supervisor to represent the communication cost.

comparatively, and therefore the norm can emerge more efficiently. However, the communication cost also increases as the cluster size becomes larger, causing higher PoM. It is obvious that the PoA and PoM are two contradictory criteria that evaluate the performance of norm emergence. PoA indicates the performance of norm emergence, while PoM indicates the cost for achieving this performance. Higher PoM indicates a more centralized system (i.e., higher cost) and thus better system performance (i.e., lower PoA). On the contrary, higher PoA indicates a more decentralized system (i.e., lower PoM) and thus lower system performance. Therefore, it is necessary to create a balance between these two criteria for the optimal performance of norm emergence. To this end, we plot the fitting curve of all the points of cluster sizes and select out the point on the curve that is closest to point (0,0) in the coordinate.[3] The point closest to (0,0), i.e., having the small value of $\sqrt{PoA^2 + PoM^2}$, can achieve the maximal system performance while bounding the administrative cost by a proper level. In the case of 30×30 grid network in Fig. 2.9, the maximal system performance can be achieved when the grid is divided into several 7×7 clusters. This point is determined using the following method: first find the tangent point on the fitting curve closest to the original point using a benchmark circle, and then find the cluster size on the fitting curve that is closest to the tangent point.

2.5 Related Work

Many studies have been done to investigate the mechanisms of norm emergence in the literature. Shoham and Tennenholtz (1997) proposed a rule based on the highest cumulative reward to study the emergence of rule. Sen and Airiau (2007) presented an emergence model in which individuals learn from another agents from the group randomly. This work is significant because it indicates that agents' private random learning is sufficient for emergence of social norms

---

[3] We use the function $f(x) = \frac{a}{(x+b)} + c$ to fit the curve, and the optimal values for coefficients a, b, and c for Fig. 2.9 are 0.0174, 0.0299, and 0.0821, respectively.

in a well-mixed agent population; Then, Sen and Airiau (2007) along with (Mukherjee *et al.*, 2008) extended this model by assessing the influence of heterogeneous agent systems and space-constrained interactions on norm emergence. Savarimuthu (Savarimuthu *et al.*, 2011; Tony and Savarimuthu, 2011) recapped the research on the multiagent-based emergence mechanisms and presented the role of three proactive learning methods (i.e., practice-based learning, observation-based learning, and communication-based learning) in accelerating norm emergence. Airiau and Sen (Airiau *et al.*, 2014; Sen and Sen, 2010) extended the group-based random learning model (Sen and Airiau, 2007) to complicated networks to study the influence of topologies on norm emergence. Villatoro *et al.* (2009) proposed a reward-learning mechanism based on interaction history. Later, they established two rules (i.e., re-building links with neighborhood and observing neighbors' behaviors) to overcome the subnorm problems (Villatoro *et al.*, 2011, 2013). Mihaylov *et al.* (2014) came up with a learning mechanism of Win–Stay Lose–Probabilistic–Shift to make 100% emergence of norms in complicated networks. Mahmoud *et al.* (2012) further extended Axelrod's seminal model by considering topological structures among agents. Yu *et al.* [2016a] proposed a series of mechanisms including the social learning strategy, the collective interaction protocol (Hao *et al.*, 2015; Yu *et al.*, 2014, 2013a), and the adaptive learning mechanism (Yu *et al.*, 2016b, 2016d) in order to promote norm emergence in networked MASs. Unlike these studies that focused on norm emergence either in a fully mixed agent population or in a network agent society, our work captures the organizational characteristic of hierarchical supervision among learning agents. This feature sets our work apart from all these existing studies.

From another perspective, the problem that has been addressed in this chapter is to employ heuristics to guide the policy search to speed up the multiagent learning process. There is plenty of work in the research of multiagent learning to properly address this problem. For example, Zhang *et al.* (2009) defined a multi-level organizational structure for automated supervision and a communication protocol for exchanging information between lower-level agents and higher-level supervising agents, and Bianchi *et al.* (2007) raised Heuristically Accelerated Minimax-Q (HAMMQ) and incorporated heuristics into the Minimax-Q algorithm to speed up convergence rate. However, HAMMQ was intended for use only in a two-agent configuration. Also, hand-coded domain heuristics are required in HAMMQ, which did not capture the dynamics of other learning agents. Our work is different from the above studies in terms of targeting a different problem (i.e., norm emergence in networked MASs). Nevertheless, the main principle embodied in the proposed framework can shed some light on understanding the learning dynamics in MAL during an efficient convergence to a predefined (i.e., Nash) equilibrium.

2.6 Conclusion

Norm emergence is a crucial issue in understanding and controlling complex systems. Assuming a fully centralized controller to govern the process of norm emergence is not only infeasible for large systems, but also expensive in terms of manageable cost. Therefore, it is more applicable for a norm to emerge on its own through agents' local interactions. This kind of pure distributed way of norm emergence, however, has another limitation in terms of low efficiency especially when the system becomes complex. The hierarchical learning framework proposed in this chapter creates a balance between centralized control and distributed interactions during the emergence of social norms, by integrating hierarchical supervision into distributed learning of agents. Experiments have shown that this compromising solution is indeed robust and efficient for evolving stable norms in networked systems, especially when the norm space is large.

As for future work, the hierarchical learning framework can be conducted in some complex networks, like small-world networks and scale-free networks. The influence of network structure on norm emergence can be explored in depth.

Acknowledgments

This work is supported by the National Natural Science Foundation of China under Grant Nos. 61502072, 61572104, and 61403059, the Hongkong Scholar Program under Grant No. XJ2017028, and Post-Doctoral Science Foundation of China under Grant Nos. 2014M561229 and 2015T80251.

Chapter 3

Making Efficient Reputation-Aware Decisions in Multiagent Systems

Han Yu*, Chunyan Miao, Bo An, Zhiqi Shen and Cyril Leung

School of Computer Science and Engineering (SCSE),
Nanyang Technological University (NTU),
Singapore 639798, Singapore

*han.yu@ntu.edu.sg

Compared to automated entities, human trustees have two distinct characteristics: (1) they are *resource constrained* (with limited time and effort to serve requests) and (2) their utility is *not linearly related* to income. Existing research in reputation-aware task delegation did not consider these two issues together. This limits their effectiveness in human–agent collectives such as crowdsourcing systems. In this chapter, we propose a distributed reputation-aware task allocation approach — RATA-NL — to address these issues simultaneously. It is designed to help an individual human trustee determine the optimal number of task requests to accept at each time step based on his situation to maximize his long-term well-being. The resulting task allocation maximizes social welfare through efficient utilization of the collective capacity of the trustees and provides provable performance guarantees. RATA-NL has been compared with five state-of-the-art approaches through extensive simulations based on human task delegation behavior abstracted from a user study involving over 100 trustees for eight weeks. The results demonstrated significant advantages of RATA-NL, especially under high workload conditions.

3.1 Introduction

In open multiagent systems (MASs) where the agents involved may be from diverse backgrounds and with potentially conflicting objectives, cooperation often needs to be built on top of reputation sanctioning mechanisms (Sen, 2013). This is inherited from human social interactions in which people have to take risks to explore how trustworthy unfamiliar interaction partners are before accumulating enough information that can help self-preservation and be shared with others (Cuesta *et al.*, 2015). In recent years, systems that involve both software agents and human beings, also known as *human–agent collectives* (HACs) (Jennings *et al.*, 2014), are starting to gain popularity. For example, in some crowdsourcing applications,[1] agent technologies are being used to help crowdsourcers (i.e., *trusters*) find suitable workers (i.e., *trustees*) to serve their task requests. In such systems, the agents play the role of mediators. Ultimately, the services are still rendered by human beings.

The emergence of HACs posts new challenges to MAS decision support research. Firstly, compared to software agents, human being are more resource constrained. A software trustee can work around the clock serving trusters' requests. However, a human trustee has more limited cognitive and physical capacity, and cannot be expected to be always available as a software trustee. This leads to the second challenge. The utility of a software trustee can be measured simply by its income. However, a human trustee's utility includes not only the income aspect, but also how additional income impacts his quality of life. In an MAS where trustees are mostly human workers (e.g., crowdsourcing), algorithmic crowdsourcing approaches — which enhance the capabilities of crowdsourcing systems by making automated calls to human expertise to solve complex problems (Yu *et al.*, 2016f) — have been proposed. Although recent works such as by Grubshtein *et al.* (2010), and Yu *et al.* (2013b, 2013d, 2015b, 2016e, 2017) are starting to address the first challenge, they are still modeling trustees' utility as linear functions with respect to their

---

[1] https://www.youtube.com/watch?v=iGlqBR5ivQ0.

income. Thus, we propose a novel new approach to address both of these challenges together.

For this purpose, the first step is to find a more realistic model of human trustees' utility function. In human factors research, *subjective well-being* (SWB) is commonly used as a holistic measure of people's quality of life (Diener, 1984). SWB has been found to increase in a nonlinear fashion with respect to increases in income (Inglehart *et al.*, 2008; Sacks *et al.*, 2010). As trustees' income increase, the rate of increase in this SWD starts to drop. Based on these observations, we propose a distributed Reputation-Aware Task Allocation approach for trustees with Non-Linear utility functions (RATA-NL). Based on the Lyapunov optimization framework (Neely, 2010), RATA-NL helps each trustee agent determine the optimal number of new task requests to accept at each time step by taking into account his current workload, eagerness to work, expected income, and maximum productivity.

We proved the existence of a *lower bound* on the ratio between the social welfare (i.e., the collective well-being of the trustee agents in an MAS) and the theoretical optimal social welfare, and an *upper bound* on the waiting time for the trusters if all trustees in an MAS follow the RATA-NL approach through queueing theory based analytical approaches. To evaluate the effectiveness of RATA-NL, we design a simulation test bed to evaluate the performance of RATA-NL against five classic and state-of-the-art approaches. The results have shown that RATA-NL significantly outperforms existing approaches, especially under conditions where the workload level in an MAS is high.

3.2 Method

In this section, we formulate the challenge of utilizing resource constrained trustee agents with nonlinear utility functions as a multi-agent optimization problem and provide an efficient solution to this problem.

3.2.1 *Problem Formulation*

In MASs, the quality and timeliness of a trustee i in serving requests are affected by his competence and resources. His performance then determines his reputation standing in the MAS. His reputation, in turn, influences trusters' future decisions on how to delegate their requests. These factors impact i's workload and income. The need for trade-off between these considerations is well documented in human factors research.

The income of a trustee i derived from performing $\mu_i(t)$ number of tasks during time step t can be expressed as

$$g(\mu_i(t)) = \sum_{j \in \mu_i(t)} u(j) \qquad (3.1)$$

where $u(j) = R_j$ if task j is completed on time with quality acceptable to the truster. Otherwise, $u(j) = 0$. R_j represents the monetary reward received by i for successfully completing task j on time. Utility functions of this form are widely used by existing works such as (Yu *et al.*, 2013b) and (Yu *et al.*, 2013d). We enrich the literature by modifying the utility function to reflect a metric most valued by human trustees according to human factors research — their SWB.

The well-being of a human trustee i is influenced by two broad categories of competing factors. On the one hand, i needs to build up his reputation to attract more work from trusters to boost his income. On the other hand, a higher workload requires more effort from i and negatively affects his general well-being. Through analyzing datasets published by multiple countries, various studies done by Diener *et al.* (1993) and Inglehart *et al.* (2008) demonstrated that the marginal increase in people's SWB generally decreases as their income increases (see Fig. 1). Recent studies to quantify the relationship between SWB and income concludes that SWB can be approximated by a function of the form of $\ln(income)$ (Sacks *et al.*, 2010). Based on these models, the SWB of i as a result of working for income during a unit time step t can be expressed as

$$\text{swb}(g(\mu_i(t))) = \ln(1 + g(\mu_i(t))) \qquad (3.2)$$

A "+1" term is included in the natural log function so that $\text{swb}(g(\mu_i(t)))$ evaluates to 0 in case $g(\mu_i(t)) = 0$. t normally

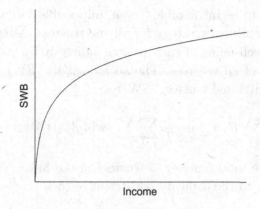

Fig. 3.1. SWB versus income.

represents a day when interpreted in the context of people working. We assume that trustees are specialized, and the tasks a trustee i is qualified to perform require similar effort from i. For example, in a crowdsourcing system, for a worker qualified to perform image labeling tasks, the effort required for him to label each image can be considered similar. Similarly, under these conditions, the value of R_j for different tasks can be regarded as the same.

The objective of this research is to design an approach to help an individual trustee i make decisions on how many of the new task requests should be accepted at any given time step t, $a_i(t)$, in order to optimize his long-term well-being. From i's perspective, to achieve this objective, he needs to complete tasks with high quality to boost his income, but not to over-work himself. In addition, this objective needs to be pursued within the framework of reputation-aware interaction in MASs. Therefore, the resulting task allocation should ensure trusters receive satisfactory quality of service in general in order to maintain a healthy level of trust in the MAS. At the same time, as trusters represent the source of income of the trustees, for the MAS to be sustainable in the long term, the trusters must be satisfied in terms of both the quality and timeliness of the service received.

Given that a large number of trustees will be involved in interactions over a potentially infinite time horizon in an uncertain environment, it is intractable (even impossible) to compute the optimal (equilibrium) strategy for all the trustees. Alternatively, we optimize the well-being of each trustee indirectly by maximizing the social welfare of all trustees. The *social welfare* (SW) is defined as the sum of individual trustees' SWB

$$\bar{U} = \lim_{T \to \infty} \frac{1}{T} \sum_{t=0}^{T-1} \sum_{i=1}^{N} \mathrm{swb}(g(\mu_i(t))) \tag{3.3}$$

where N is the total number of trustees in the MAS. The optimization is subject to the capacity constraint of each trustee

$$0 \leq \mu_i(t) \leq \mu_i^{\max}, \quad \forall i, t \tag{3.4}$$

where $\mu_i^{\max}$ denotes the maximum number of tasks i can process per unit time. It is reasonable to assume that an agent of i can obtain the value of $\mu_i^{\max}$ through observing i's past performance. The $\mu_i^{\max}$ value can also be obtained by letting i provide an estimation.

Maximizing the sum of nonlinear utility functions is a challenging problem, especially in a distributed fashion. To make the optimization problem more tractable, we view the MAS as a queueing system and analyze the optimization under the framework of reputation-aware decision making. The queueing dynamics of a trustee i's pending task queue at time $t + 1$, $q_i(t + 1)$, can be expressed as

$$q_i(t+1) = q_i(t) - \mu_i(t) + a_i(t) \tag{3.5}$$

where $a_i(t)$ is the number of new tasks an agent i should accept at time t.

As trustees are resource-constrained entities, the MAS can be shown to resemble a congestion game environment (Yu *et al.*, 2013c) where the quality of service received by trusters is affected by how tasks are distributed among trustees.

To help human trustees optimize their well-being (i.e., the time-averaged value of Eq. (3.2)) when working in an MAS, a trustee agent that can help its owner determine the optimal number of new requests to be accepted in real time with a balanced consideration of his income and his reputation among the trusters is needed.

3.2.2 An Efficient Distributed Decision-Making Approach

The RATA-NL approach is designed for individual trustees and requires only local knowledge about the trustee as input. Nevertheless, it balances the benefits for both parties involved in an interaction — the truster and the trustee. Intuitively, the task acceptance decision made by RATA-NL at any given point in time can be summarized as follow: "the more *eager to work* a trustee i is, the lighter his *current workload*, and the larger the *expected reward* for completing a task, the more new task requests i should accept." In this section, we present how such an intuition can be translated into an actionable task acceptance decision approach for trustees by combining reputation-aware decision-making with queueing theory.

Let $\vec{q}(t) = (q_i(t))$ be a vector denoting the lengths of pending task queues of all trustees in an MAS. A commonly used metric for measuring the overall level of congestion in queueing system is the quadratic *Lyapunov function* (Neely, 2010) in the form of $L(\vec{q}(t)) = \frac{1}{2}\sum_{i=1}^{N}(q_i(t))^2$. Here, the coefficient $\frac{1}{2}$ is added to simplify the notations in subsequent analysis. We adopt this metric to measure the overall level of congestion in an MAS. Under the same framework, the stepwise change in the congestion level in the MAS can be measured by the *conditional Lyapunov drift* (Neely, 2010), $\triangle(\vec{q}(t))$, which is expressed as

$$\triangle(\vec{q}(t)) = \sum_{i=1}^{N} \mathbb{E}\{L(\vec{q}(t+1)) - L(\vec{q}(t))|\vec{q}(t)\} \tag{3.6}$$

where the conditional expectation is with respect to the one-step queueing dynamics given the current $\vec{q}(t)$. To ensure the decisions made by RATA-NL will not result in indefinite build-up of pending tasks in any $q_i(t) \in \vec{q}(t)$, $\triangle(\vec{q}(t))$ needs to be minimized.

By squaring both sides of Eq. (3.5), we have:

$$q_i^2(t+1) = q_i^2(t) + 2q_i(t)[a_i(t) - \mu_i(t)] + a_i^2(t) + \mu_i^2(t) - 2a_i(t)\mu_i(t) \tag{3.7}$$

By re-arranging the terms and dividing both sides by 2

$$\frac{1}{2}q_i^2(t+1) - \frac{1}{2}q_i^2(t) \leq \frac{1}{2}\mu_i^2(t) + \frac{1}{2}a_i^2(t) - q_i(t)[\mu_i(t) - a_i(t)] \quad (3.8)$$

Due to the capacity constraints of the trustees, there exists a constant $\mu_i^{\max} \geq \mu_i(t)$. To minimize the risk of i being overloaded with work, $a_i(t)$ values should be selected such that it is also less than or equal to $\mu_i^{\max}$. Thus, by taking expectation on both sides of Eq. (3.8), we have:

$$\triangle(\vec{q}(t)) \leq \sum_{i=1}^{N} [(\mu_i^{\max})^2 - q_i(t)\mathbb{E}\{\mu_i(t) - a_i(t)|q_i(t)\}] \quad (3.9)$$

When determining $a_i(t)$, it is necessary to estimate its potential impact on i's well-being in the subsequent time step (i.e., $\text{swb}(g(\mu_i(t+1)))$). As the distribution of $\mu_i(t+1)$ depends on the actual performance of i and cannot be known in advance, $\text{swb}(g(\mu_i(t+1)))$ cannot be directly computed. Thus, we propose an alternative way to estimate $\text{swb}(g(\mu_i(t+1)))$ based on $a_i(t)$. The expected income from accepting $a_i(t)$ number of new task requests can be expressed in a similar way as in (Yu *et al.*, 2013b):

$$\hat{g}(a_i(t)) = \mathbb{E}\left\{\sum_{j \in a_i(t)} u(j)\right\} = r_i(t)a_i(t)R_i \quad (3.10)$$

where R_i is the reward received by i, on average, for a task successfully completed on time. $r_i(t)$ is i's current reputation. $r_i(t)$ can be computed with any existing reputation model as long as the value produced by the model is within the range of $[0, 1]$. This enables $r_i(t)$ to be interpreted as the probability for i to complete a task on time with high quality at t. In many MAS-like systems such as an e-commerce website, the reputation of a seller (i.e., a trustee) is public knowledge. Thus, it is reasonable to assume that the trustee knows this value, too.

In order to ensure the stability of the pending task queues at all trustees (i.e., none of the pending task queues will grow in length indefinitely), the time averaged task acceptance rate must not exceed the time-averaged task processing rate for all i (Neely, 2010):

$$\lim_{T \to \infty} \frac{1}{T} \sum_{t=0}^{T-1} a_i(t) \leq \lim_{T \to \infty} \frac{1}{T} \sum_{t=0}^{T-1} \mu_i(t) \tag{3.11}$$

Each individual trustee i uses RATA-NL to determine the value of $a_i(t)$ at each time step to maximize i's SWB. At the same time, the collective decisions by trustees need minimize the general level of congestion in the MAS so that its long-term operation can be sustained. From the perspective of a given MAS, these two considerations can be expressed as a *congestion-minus-well-being* expression

$$\Delta(\vec{q}(t)) - \sum_{i=1}^{N} \Psi_i \mathbb{E}\{\text{swb}(g(\mu_i(t)))|q_i(t)\} \tag{3.12}$$

which needs to be minimized. $\Psi_i > 0$ is a control parameter to be specified by the trustee indicating his eagerness to work. The larger the value of Ψ_i, the more *eager to work* i is. Since swb($\cdot$) is a concave and monotonically increasing function, based on Eq. (3.11), we have:

$$\mathbb{E}\{\text{swb}(g(\mu_i(t)))|q_i(t)\} \geq \mathbb{E}\{\text{swb}(\hat{g}(a_i(t)))|q_i(t)\} \tag{3.13}$$

Therefore, based on Eqs. (3.9) and (3.13), Eq. (3.12) satisfies

$$\Delta(\vec{q}(t)) - \sum_{i=1}^{N} \Psi_i \mathbb{E}\{\text{swb}(g(\mu_i(t)))|q_i(t)\}$$

$$\leq \Delta(\vec{q}(t)) - \sum_{i=1}^{N} \Psi_i \mathbb{E}\{\text{swb}(\hat{g}(a_i(t)))|q_i(t)\}$$

$$\leq \delta - \sum_{i=1}^{N} \Psi_i \mathbb{E}\{\text{swb}(\hat{g}(a_i(t)))|q_i(t)\}$$

$$- \sum_{i=1}^{N} q_i(t) \mathbb{E}\{\mu_i(t) - a_i(t)|q_i(t)\}$$

where N is the total number of trustees in the MAS. To simplify the notations, we define a constant $\delta = \sum_{i=1}^{N} (\mu_i^{\max})^2$.

As $\mu_i(t)$ is not controlled by the RATA-NL approach, it can be excluded from the objective function. By isolating terms containing $a_i(t)$ on the right-hand side of the above inequality, we have

$$\Psi_i \text{swb}(\hat{g}(a_i(t)))|q_i(t)\} - \sum_{i=1}^{N} \mathbb{E}\{q_i(t)a_i(t)\} \tag{3.14}$$

Thus, in order to maximize $\sum_{i=1}^{N} \Psi_i \mathbb{E}\{\text{swb}(g(\mu_i(t)))|q_i(t)\} - \Delta(\bar{q}(t))$, Eq. (3.14) needs to be maximized by selecting appropriate values of $a_i(t)$ at each time step. The objective function is to maximize the expected SWB while minimizing fluctuations in agents' workload. It is similar to a recent finding in human choice behavior suggestion we make decisions based on a value maximization and surprise minimization heuristic (Schwartenbeck *et al.*, 2015).

Maximize:

$$Obj_i(t) = \Psi_i \ln(1 + r_i(t)a_i(t)R_i) - q_i(t)a_i(t) \tag{3.15}$$

Subject to:

$$0 \le a_i(t) \le \mu_i^{\max} \tag{3.16}$$

$$0 \le a_i(t) \le \lambda_i(t), \forall t \tag{3.17}$$

where $\lambda_i(t)$ is the number of task requests for i at time step t. Its value depends on the decision-making process of the trusters, which is influenced by the trustees' reputation values. The value of $a_i(t)$ can be solved by differentiating Eq. (3.15), which is a convex function, with respect to $a_i(t)$ and finding the critical point subject to Constraints (3.16) and (3.17)

$$\frac{\partial Obj_i(t)}{\partial a_i(t)} = \frac{\Psi_i r_i(t)R_i}{1 + r_i(t)a_i(t)R_i} - q_i(t) = 0 \tag{3.18}$$

$$a_i(t) = \min\left[\max\left[\left\lfloor \frac{\Psi_i}{q_i(t)} - \frac{1}{r_i(t)R_i}\right\rfloor, 0\right], \mu_i^{\max}, \lambda_i(t)\right] \tag{3.19}$$

Equation (3.19) can be interpreted as the following task acceptance policy: "the more *eager to work* a trustee i is (indicated by large Ψ_i

values), the lighter his *current workload*, and the larger the *expected reward* for completing a task, the more new task requests i should accept subject to his physical limitation (Constraint (3.16)) and the actual number of task requests directed at him (Constraint (3.17))". Such a policy is rational for a human trustee and provides actionable guidance for a software agent of the trustee to compute the exact value of $a_i(t)$.

The RATA-NL approach is listed in Algorithm 3.1. It is designed for usage by individual trustees in a distributed fashion. In the case where not all incoming task requests are accepted by i, the RATA-NL approach informs the requesting trusters so that they can look for other alternatives. Throughout this process, no communication among trustees is required. The input for the variables required by RATA-NL can reasonably be assumed to be available with proper monitoring mechanisms in an MAS. A trustee only needs to provide a value for Ψ_i to RATA-NL following guidelines, to be discussed in Section 3.3.1.

3.3 Results

3.3.1 *Theoretical Analysis*

In this section, we analyze the impact on the social welfare of an MAS and the waiting time experienced by the trusters if RATA-NL were to be adopted by all trustees.

Algorithm 3.1 RATA-NL
___
Require: $q_i(t)$, $\lambda_i(t)$, $r_i(t)$, Ψ_i, R_i and $\mu_i^{\max}$.
1: **if** $\lambda_i(t) > 0$ **then**
2:    **if** $q_i(t) = 0$ **then**
3:        $a_i(t) = \min[\mu_i^{\max}, \lambda_i(t)]$;
4:    **else**
5:        $a_i(t) = \min[\max[\lfloor \frac{\Psi_i}{q_i(t)} - \frac{1}{r_i(t)R_i} \rfloor, 0], \mu_i^{\max}, \lambda_i(t)]$;
6:    **if** $\lambda_i(t) - a_i(t) > 0$ **then**
7:        Return $\lambda_i(t) - a_i(t)$ unaccepted tasks to requesters;
8: **else**
9:    $a_i(t) = 0$
___

Let $U^*(t)$ be the theoretical optimal social welfare produced by an MAS at t based on perfect foresight. Assume there are positive values Ψ, δ, and ϵ such that the *congestion-minus-utility* expression in Eq. (3.12) satisfies

$$\Delta(\vec{q}(t)) - \Psi \sum_{i=1}^{N} \mathbb{E}\{\text{swb}(g(\mu_i(t)))|q_i(t)\} \leq \delta - \epsilon \sum_{i=1}^{N} q_i(t) - \Psi U^*(t)$$

where $\Psi = \frac{1}{N} \sum_{i=1}^{N} \Psi_i$. Taking expectations over the distribution of all $q_i(t)$ on both sides, we have

$$\sum_{i=1}^{N} \mathbb{E}\{L(q_i(t+1)) - L(q_i(t))|q_i(t)\} - \Psi \sum_{i=1}^{N} \mathbb{E}\{\text{swb}(\mu_i(t))|q_i(t)\}$$

$$\leq \delta - \epsilon \sum_{i=1}^{N} \mathbb{E}\{q_i(t)\} - \Psi U^*(t)$$

which holds for all time steps. Summing both sides of the above expression over $t \in \{0, 1, \ldots, T-1\}$ yields

$$\sum_{t=0}^{T-1} \sum_{i=1}^{N} \mathbb{E}\{L(q_i(t+1)) - L(q_i(t))\} - \Psi \sum_{t=0}^{T-1} \sum_{i=1}^{N} \mathbb{E}\{\text{swb}(\mu_i(t))\}$$

$$= \sum_{i=1}^{N} \mathbb{E}\{L(q_i(T)) - L(q_i(0))\} - \Psi \sum_{t=0}^{T-1} \sum_{i=1}^{N} \mathbb{E}\{\text{swb}(\mu_i(t))\}$$

$$\leq T\delta - \epsilon \sum_{t=0}^{T-1} \sum_{i=1}^{N} \mathbb{E}\{q_i(t)\} - \Psi \sum_{t=0}^{T-1} U^*(t)$$

By re-arranging the terms in the above inequality, we have

$$\epsilon \sum_{t=0}^{T-1} \sum_{i=1}^{N} \mathbb{E}\{q_i(t)\} \leq T\delta + \Psi \sum_{t=0}^{T-1} \sum_{i=1}^{N} \mathbb{E}\{\text{swb}(\mu_i(t))\}$$

$$- \sum_{t=0}^{T-1} U^*(t) - \sum_{i=1}^{N} \mathbb{E}\{L(q_i(T))\} + \sum_{i=1}^{N} \mathbb{E}\{L(q_i(0))\}$$

Since $U^*(t) > 0$, $L(\cdot) \geq 0$ and $L(q_i(0)) = 0$, the above inequality can be simplified as:

$$\epsilon \sum_{t=0}^{T-1} \sum_{i=1}^{N} \mathbb{E}\{q_i(t)\} \leq T\delta + \Psi \sum_{t=0}^{T-1} \sum_{i=1}^{N} \mathbb{E}\{\text{swb}(\mu_i(t))\}$$

Let $\text{swb}_{\max} = \max_{[\forall n,t]} \text{swb}(\hat{g}(a_i(t)))$ be the largest observed per time step utility of any trustee in the MAS up to $T - 1$ such that $\Psi \sum_{t=0}^{T-1} \sum_{i=1}^{N} \mathbb{E}\{\text{swb}(\mu_i(t))\} \leq NT\Psi\text{swb}_{\max}$. The above inequality can be written as:

$$\epsilon \sum_{t=0}^{T-1} \sum_{i=1}^{N} \mathbb{E}\{q_i(t)\} \leq T(\delta + N\Psi\text{swb}_{\max})$$

By dividing both sides by $T\epsilon$, the *upper bound* on the time-averaged task queue lengths for all trustees in an MAS is

$$\limsup_{T \to \infty} \frac{1}{T} \sum_{t=0}^{T-1} \sum_{i=1}^{N} \mathbb{E}\{q_i(t)\} \leq \frac{\delta + N\Psi\text{swb}_{\max}}{\epsilon}$$

Similarly, the *lower bound* on the time-averaged social welfare produced by an MAS is

$$\liminf_{T \to \infty} \frac{1}{T} \sum_{t=0}^{T-1} \sum_{i=1}^{N} \mathbb{E}\{\text{swb}(g(\mu_i(t)))\}$$

$$\geq \frac{1}{T} \sum_{t=0}^{T-1} U^*(t) - \frac{\delta}{\Psi} - \frac{\sum_{i=1}^{N} \mathbb{E}\{L(q_i(0))\}}{T\Psi}$$

$$+ \frac{\epsilon \sum_{t=0}^{T-1} \sum_{i=1}^{N} \mathbb{E}\{q_i(t)\} + \sum_{i=1}^{N} \mathbb{E}\{L(q_i(T))\}}{T\Psi}$$

$$\geq \frac{1}{T} \sum_{t=0}^{T-1} U^*(t) - \frac{\delta}{\Psi}$$

Therefore, under the condition where the task allocation recommendations by RATA-NL to all trustees are fully complied with, and assuming the reputation values produced by the underlying reputation model correctly reflect the trustees' behavior, the MAS can produce time-averaged social welfare within $O(1/\Psi)$ of the optimal social welfare with average waiting time experienced by the trusters bounded by $O(\Psi)$.

3.3.2 *Simulations*

To evaluate the performance of RATA-NL, we build a simulation test bed with diverse behavior patterns and compare the performance of RATA-NL with five other state-of-the-art approaches.

3.3.2.1 *Experiment settings*

In the simulation test bed, we assume binary outcomes for task results (i.e., a task is considered successfully completed by a trustee agent if the quality of the result is satisfactory and the result is produced before its expected deadline. Otherwise, it is considered unsuccessful). We create six groups of 100 trustee agents each to compare the performance of six different approaches. They are as follows:

(1) The Equality-based Approach (EA): this is an approach based on the patterns exhibited by participants in the EA Group in our dataset.
(2) The Trust-based Approach (TA): this is also an approach based on the patterns exhibited by participants in the TA Group in our dataset. Each truster agent uses its direct interaction experience with trustees in the past to evaluate their trustworthiness using the *BRSEXT* method described by Yu *et al.* (2013c). At each time step, a truster agent delegates tasks to the most trustworthy known trustee.
(3) The Reputation-based Approach (RA): each truster agent then adopts the approach by Vogiatzis *et al.* (2010) in which the probability of a trustee agent being selected by a truster agent is directly related to its reputation standing among all trustee agents in the MAS.
(4) The Global Considerations (GC) Approach: truster agents adapt the probability for each trustee agent being selected to

serve task requests following the approach by Grubshtein *et al.* (2010).

(5) The DRAFT Approach: trustee agents make request acceptance decisions following the approach by Yu *et al.* (2013b).

(6) The RATA-NL Approach: trustee agents make request acceptance decisions following the proposed approach.

Truster agents in Approaches 4–6 also adopt the approach by Vogiatzis *et al.* (2010).

Trustee agents behave following one of the listed patterns:

(1) *Com*: Competent trustees who produce satisfactory quality results with a 90% probability;

(2) *MC*: Moderately competent trustees who produce satisfactory quality results with a 70% probability;

(3) *MI*: Moderately incompetent trustees who produce satisfactory quality results with a 30% probability;

(4) *Inc*: Incompetent trustees who produce satisfactory quality results with a 10% probability.

The task processing capacities ($\mu_i^{\max}$ values) of each type of trustee agents are set in such a way that more competent agents can process more tasks per time step than less competent ones. In all simulations, we set $\Psi_i = \mu_i^{\max}$ When we refer to a trustee agent population as "$x\%$ competent", the exact composition of the population consists of $\frac{1}{2}x\%$ *Com* trustees, $\frac{1}{2}x\%$ *MC* trustees, $\frac{1}{2}(100 - x)\%$ *MI* trustees, and $\frac{1}{2}(100 - x)\%$ *Inc* trustees. In the experiments, the trustee agent population compositions are varied from 10% to 100% competent to simulate different trustee behavior patterns.

Another factor affecting the well-being of trustees is the general level of workload in an MAS. As the workload is relative to the aggregate task processing capacity of a given trustee population, we define a formula to compute the *maximum throughput*, θ, of a trustee population per time step as $\theta = \sum_{i=1}^{N} c_i \mu_i^{\max}$, where c_i denotes the competence value of a trustee i. The workload on a given trustee population is measured by a metric called *Load Factor* (LF) which is computed as $LF = \frac{N_{\text{req}}}{\theta}$, where N_{req} is the average number of task requests generated by trusters per time step. In the experiments, we vary the *LF* value from 25% to 200% to simulate different workload conditions. Under each configuration, the simulation is run for $T = 1000$ time steps. In all experiments, trustees process tasks at an average rate of $0.9\mu_i^{\max}$ with a standard deviation of $0.1\mu_i^{\max}$. On average, trusters expect a task to be completed within two time steps after it is allocated to a trustee.

In the experiments, we measure the performance of each approach using the following metrics:

(1) *Social welfare to optimal social welfare ratio* (SW/Opt SW): The optimal social welfare (Opt SW), U^*, that can be produced by a trustee population per time step is expressed as $U^* = \sum_{i=1}^{\lfloor \min[1.0, LF] N \rfloor} \ln(1 + c_i \mu_i^{\max} R_i)$. The N trustee agents are sorted in descending order of their c_i values. U^* can only be calculated in a controlled experimental environment where LF, $\mu_i^{\max}$, c_i and R_i can be definitively known. The SW/Opt SW ratio is calculated as $\bar{U}/U^*$.

(2) *High quality rate* (HQR): This metric is computed as $\frac{N_q}{N_{\mathrm{acc}}}$ where N_q denotes the total number of tasks completed with satisfactory quality, and N_{acc} denotes the total number of tasks accepted by the trustees over T time steps in a simulation.

(3) *Timely completion rate* (TCR): This metric is computed as $\frac{N_t}{N_{\mathrm{acc}}}$ where N_t denotes the total number of tasks completed before the expected deadlines over T time steps in a simulation.

The higher the values for these metrics, the better the performance of an approach.

In addition, we also measure how different approaches may affect the truster agents' perceptions on the behavior of the trustees. It is important as it will affect the truster agents' subsequent task delegation decisions and, in turn, the trustee agents' well-being. Ideally, the reputation value of a trustee should only reflect its competence rather than performance variations caused by changing workloads which is not the trustee's own fault. We adopt commonly used metrics including *precision*, *recall*, *f-value*, and *mean absolute error* (MAE) to measure how accurately each approach classifies whether trustees are trustworthy against the ground truth. Precision=$\frac{t_p}{t_p+f_p}$, Recall=$\frac{t_p}{t_p+f_n}$, and f-value=$\frac{2\times\text{Precision}\times\text{Recall}}{\text{Precision}+\text{Recall}}$, where t_p (*true positive*) is the number of trustees correctly classified as competent (i.e., $t_p = \frac{1}{T}\sum_{t=0}^{T-1}\sum_{i=1}^{N}1_{[r_i(t)>0.5|c_i>0.5]}$), f_p (*false positive*) is the number of trustees incorrectly classified as competent (i.e., $f_p = \frac{1}{T}\sum_{t=0}^{T-1}\sum_{i=1}^{N}1_{[r_i(t)>0.5|c_i\leq0.5]}$), and f_n (*false negative*) is the number of trustees incorrectly classified as incompetent (i.e., $f_n = \frac{1}{T}\sum_{t=0}^{T-1}\sum_{i=1}^{N}1_{[r_i(t)\leq0.5|c_i>0.5]}$). $1_{[\text{condition}]}$ equals to 1 if [condition] is true, and 0 otherwise. MAE is defined as $\frac{1}{TN}\sum_{t=0}^{T-1}\sum_{i=1}^{N}|r_i(t) - c_i|$.

3.3.2.2 *Simulation results*

Figure 3.2 contains subfigures showing the performance of the various approaches according to the evaluation metrics. Each data point in these figures represents the average value of the selected metric taken over 10 different trustee population configurations (10–100% competent) under a given load factor. As the results are based on simulations, the trends and relative performances of the approaches are more important than the exact numerical values.

The differences in performance among various approaches measured by their recall values are more significant (Fig. 3.2(a)). With $LF < 1.0$, under EA, TA, RA, GC, and RATA-NL, most unsuccessful tasks are caused by the competence of the trustees. Only a negligible percent of unsuccessful tasks are caused by failure to be completed on time due to trustees being overloaded. With $LF \geq 1.0$, the f_n values for EA, TA, RA, GC, and RATA-NL start to increase as more trustees suffer from reputation damage due to overloading, causing the recall values of these approaches to decrease. The stringent task allocation criteria used by DRAFT resulted in over-concentration of tasks on reputable trustees under low workload conditions. As workload becomes higher, the performance of DRAFT improves to close to that of RATA-NL. These factors result in the relative performances of the approaches measured with their f-values (Fig. 3.2(b)) and MAE values (Fig. 3.2(c)). In terms of accurately reflecting the behavior of the trustees rather than the dynamics caused by inefficient task

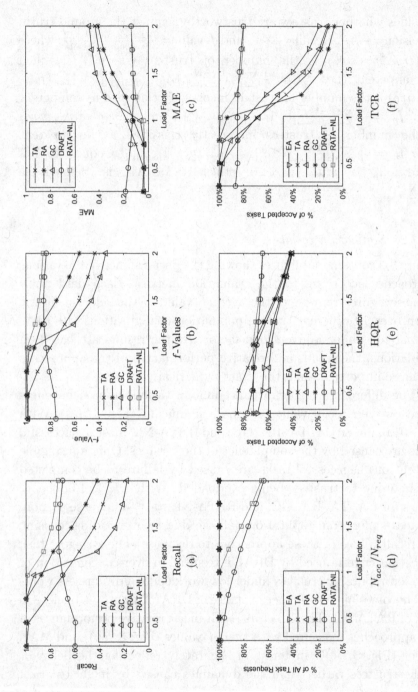

Fig. 3.2.   Comparison of performance.

allocation decisions, RATA-NL significantly outperforms EA, TA, RA, and GC under high workload conditions, and significantly outperforms DRAFT under low workload conditions.

Figure 3.2(d) shows the ratio between the accepted tasks and all proposed tasks under different approaches. Since EA, TA, RA, and GC do not provide mechanisms for trustees to reject incoming task requests, regardless of how the LF value changes, all proposed tasks are delegated to some trustees for processing. Under DRAFT, N_{acc} starts to drop even when workload is low ($LF < 1.0$). In the case of RATA-NL, all proposed tasks are allocated to trustees when $LF \leq$ 1.0 and N_{acc} only starts to drop when $LF > 1.0$. When $LF > 1.0$, the task request arrival rates become larger than the task processing rates the trustee populations can effectively support. In this case, if the extra tasks are not dropped, they can cause delays and negatively affect the performance of the trustees as perceived by the trusters. Under such conditions, to protect the long term well-being of the trustees, it is advantageous to drop some requests.

Under $LF < 1.0$, the HQR achieved by TA beats other approaches (Fig. 3.2(e)). However, as LF increases, the HQRs achieved by EA, TA, RA, and GC dropped significantly (for TA, as much as 45 percentage points), whereas for DRAFT and RATA-NL, the HQRs remains relatively stable under changing workload conditions. A similar performance pattern can be observed from their TCRs (Fig. 3.2(f)). Under this metric, the performance of RATA-NL matches that of DRAFT with $LF > 1.0$.

With complex interactions among the above-mentioned factors, the overall performance landscape, in terms of SW/Opt SW, for each of the six approaches under different workload and trustee trustworthiness distributions are shown in Fig. 3.3. EA achieves the best performance under low to medium workload with highly trustworthy trustees. When load factor exceeds 1.0, the more highly trustee agents with low productivity are present, the worse the performance of EA. The performance landscape of TA is very similar to EA, but with lower SW/Opt SW. The same can be said for RA. However, the average SW/Opt SW achieved by RA is higher than that achieved by TA. GC achieves the highest performance under

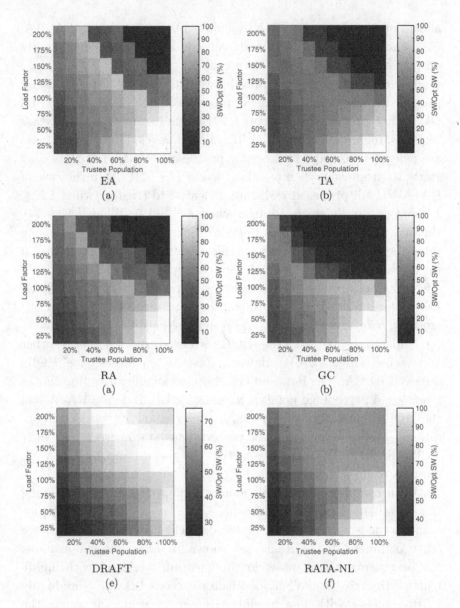

Fig. 3.3. Performance landscape of comparison approaches.

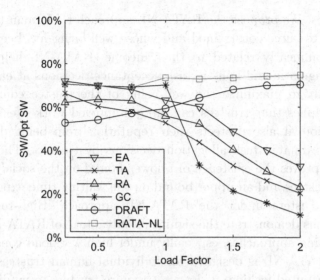

Fig. 3.4. SW/Opt SW.

$LF \leq 1.0$, but the worst performance once the overall workload exceeds 1.0. DRAFT achieves the worst performance for $LF \leq 1.0$, but much higher performance compared to EA, TA, RA, and GC under $LF > 1.0$. RATA-NL achieves comparable performance with GC under $LF \leq 1.0$, and slightly better performance than DRAFT under $LF > 1.0$.

Figure 3.4 shows the SW/Opt SW ratios achieved by various approaches. With $LF \leq 1.0$, the differences in the performance of various approaches are not large (within 15%). The performance of RATA-NL, GC, and EA are almost the same, with RA and TA trailing not far behind. DRAFT delivered the worst performance under low LF conditions where the capacity of the trustees cannot be fully utilized. With $LF > 1.0$, the advantage of RATA-NL and DRAFT over other approaches become more significant with RATA-NL outperforming DRAFT by about 10%.

3.4 Discussions

In this chapter, we take a step toward bringing the discussion about reputation-aware decision-making into the domain of human–agent collectives. We propose the RATA-NL approach for human trustees, who are resource constrained and whose well-being has been found to be nonlinearly related to their income. RATA-NL helps them make pragmatic and holistic task acceptance decisions at each time step so as to maximize the well-being of the trustee community, which implies that trustees can enjoy improved work–life balance. In addition, it also protects their reputation from being damaged by uncoordinated task allocation decisions by trusters. Theoretical analysis proves the existence of a lower bound on the social welfare of the trustees and an upper bound on the waiting time experienced by the trusters under the RATA-NL approach. High-resolution simulations demonstrate the significant advantage of RATA-NL over five existing approaches, especially under high workload conditions.

As RATA-NL is designed for individual human trustees to use in a distributed fashion, it does not require trusters to modify their existing decision-making process. This makes it attractive for use in deployed systems such as e-commerce platforms to enhance user experience and overall system performance. In order to enable agents to operate in an MAS together with human beings, more data are needed to construct models of human performance and decision-making behaviors. In future research, we plan to set up a game-based online platform for long-term collection of such data to build a benchmark for this research topic.

Acknowledgments

This research is supported by the National Research Foundation, Prime Minister's Office, Singapore under its IDM Futures Funding Initiative; and the Lee Kuan Yew Post-Doctoral Fellowship Grant. This book chapter has been extended from the conference paper: Yu, H., Miao, C., An, B., Shen, Z. & Leung, C. Reputation-aware task allocation for human trustees. In *Proceedings of the 13th International Conference on Autonomous Agents and Multi-Agent Systems (AAMAS'14)*, pp. 357–364 (2014).

Chapter 4

Decision-Theoretic Planning in Partially Observable Environments

Zongzhang Zhang[*,‡] and Mykel Kochenderfer[†,§]

*School of Computer Science and Technology,
Soochow University,
Suzhou 215006, China

†School of Engineering, Aeronautics and Astronautics,
Stanford University,
Stanford CA 94305-4035, USA

‡zzzhang@suda.edu.cn
§mykel@stanford.edu

In environments where the effects of actions are stochastic and observations incomplete, partially observable Markov decision processes (POMDPs) serve as a powerful mathematical model for single-agent sequential decision making. However, solving POMDPs exactly is challenging in practice due to their computational complexity. Over the past 15 years, there have been significant advances in scalable approximate POMDP planning methods in both offline and online settings, allowing for important real-world applications. Finding optimal or near-optimal solutions to POMDPs is often challenging due to two distinct but related reasons: the curse of dimensionality, where the computation is directly related to the number of states, and the curse of history, where the number of distinct histories grows exponentially as the planning horizon increases. On the theoretical front, the covering number, a metric of capturing the curse of history, has been proposed as a complexity measure for approximate POMDP planning. This chapter begins by introducing the POMDP model, offline and online POMDP solution methods, followed by a description of theoretical results based on covering numbers.

4.1 Introduction

This chapter focuses on the computational issues involved in making optimal sequential decisions in uncertain environments. Two kinds of uncertainty can make decision making challenging: (1) the actions can have stochastic effects and (2) the state of the environment is not known exactly due to imperfect sensors, i.e., the environment is only partially observable. The partially observable Markov decision process (POMDP) provides a rich mathematical framework for sequential decision-making problems under these kinds of uncertainty (Kaelbling *et al.*, 1998; Kochenderfer, 2015). The framework originated in the operations research community and was later adopted by the AI and automated planning communities. The POMDP model can be used in a wide range of applications such as object grasping (Hsiao *et al.*, 2007), mobile robot exploration (Pineau *et al.*, 2006a; Spaan and Vlassis, 2004), spoken dialogue management (Thomson and Young, 2010), automated handwashing assistance (Hoey *et al.*, 2010), automated fault recovery (Shani and Meek, 2009), unmanned aircraft collision avoidance (Bai *et al.*, 2011; Wolf and Kochenderfer, 2011), and many others (Grześ *et al.*, 2015; Hauskrecht, 2000; White III and Cheong, 2012).

Exact planning in POMDPs (Cassandra *et al.*, 1997; Kaelbling *et al.*, 1998; Zhang and Zhang, 2001) involves computing an optimal plan for selecting actions from all possible information states (also known as beliefs). These algorithms suffer from the well-known curse of dimensionality, where the computational complexity of the planning problem is directly related to the number of states. In addition, these algorithms suffer from the less known curse of history, where the number of distinct histories grows exponentially as the planning horizon increases. As a result, solving a POMDP optimally is often computationally intractable except for small toy problems with a handful of states (Shani *et al.*, 2013). Hence, there is tremendous interest in approximate solution techniques.

Over the past 15 years, approximate algorithms have received attention because of their ability to compute successful policies for increasingly large problems. Point-based value iteration methods (Kurniawati *et al.*, 2008; Pineau *et al.*, 2003, 2006a; Poupart *et al.*, 2011; Shani *et al.*, 2013; Smith and Simmons, 2005; Zhang *et al.*, 2014, 2015) are a family of approximate algorithms that have been the focus of recent research. These methods compute solutions only for a finite subset of beliefs in the belief space. Currently, point-based solvers are capable of handling complex domains with many thousands of states in a matter of seconds by exploring only the reachable belief space, i.e., a subset of the belief space that can be encountered by interacting with the environment. All these methods are offline algorithms, meaning that they, prior to execution, perform all or most of the computation required for determining the best action to execute for all possible belief states.

In large POMDP planning tasks, a viable alternative may be to use an online approach (Bai *et al.*, 2014; Ross *et al.*, 2008b; Silver and Veness, 2010; Somani *et al.*, 2013; Zhang and Chen, 2012), which computes a local policy only for the current belief. Online approaches alternate between a time-limited policy-construction step for the current state and a policy-execution step (Ross *et al.*, 2008b). Such an approach can potentially produce a sequence of actions with high return while spending much less time overall in policy construction and policy execution compared to offline algorithms.

It may seem surprising that even an approximate solution can be obtained in seconds in a space of thousands of dimensions. POMDP researchers have studied some of the reasons why these point-based algorithms work so well (Hsu *et al.*, 2007; Pineau *et al.*, 2006a). They found that the success of these algorithms is due to the fact that the curse of history often plays a more important role in affecting POMDP value iteration than the curse of dimensionality. The covering number of the reachable belief space provides a metric for capturing the curse of history, and therefore can be used to quantify the complexity of approximate POMDP planning (Hsu *et al.*, 2007; Zhang *et al.*, 2012). Intuitively, the covering number of a space B is the minimum number of fixed-size balls required to cover the space B so that all points in B lie within some ball. Recent work (Zhang *et al.*, 2014) has used the covering number to characterize the size of the search space reachable under heuristic policies and connect the complexity of POMDP planning to the effectiveness of heuristics.

The chapter is organized as follows. First, Section 4.2 describes the basic concepts in the POMDP model, including the Markov decision process (MDP), belief, belief updating, the belief-state MDP model, and policy and value function representations. Next, Section 4.3 reviews offline solution techniques for POMDPs, emphasizing exact value iteration and point-based value iteration methods. Section 4.4 gives an overview of online planning techniques, highlighting branch-and-bound search, heuristic search, and Monte Carlo tree search methods. Section 4.5 describes the main theoretical results that connect the covering number concept to the complexity of approximate POMDP planning. We end with concluding remarks in Section 4.6.

4.2 Partially Observable Markov Decision Processes

This section begins by describing the MDP model and then its generalization, the POMDP model. We will consider only finite models, which are the most commonly used in the POMDP literature given the difficulties of solving continuous models.

4.2.1 *MDPs*

The Markov decision process (MDP) models sequential decision problems under uncertainty about the effect of an agent's action. The environment is assumed fully observable (Kochenderfer, 2015; Sutton and Barto, 1998a; Wiering and van Ottelo, 2012). This assumption is relaxed in the POMDP framework.

An MDP can be defined as a tuple $\langle S, A, T, R, \gamma \rangle$:

- S is a finite set of states in which a decision can be made;
- A is a finite set of actions;

- $T : S \times A \to \Pi(S)$ is the state-transition function, where $T(s' \mid s, a)$ denotes the probability of transitioning from s to s' after taking action a;
- $R : S \times A \to \mathbb{R}$ is the reward function, where $R(s, a)$ denotes the expected immediate reward received for taking action a in state s;
- $\gamma \in [0, 1]$ is the discount factor.

Such an MDP environment has the Markov property, i.e., the next state and the expected reward depend only on the current state and action, and not on any previous state or action. MDPs are an extension of Markov chains; the difference is the addition of actions and rewards to allow for goal-directed behaviors.

The goal for any given MDP is to act optimally to maximize the expected discounted future reward $E[\sum_{t=0}^{\infty} \gamma^t r_t]$, where γ represents the difference in importance between immediate and future rewards. When $\gamma = 0$, the agent only cares about which actions will yield the maximum expected immediate reward. As γ approaches 1, the agent becomes more farsighted, i.e., it takes future rewards into account more strongly.

Figure 4.1 shows an MDP agent taking the state and reward as input and interacting with the environment by following a policy. In an MDP, a policy is a state–action mapping defined as $\pi : S \to A$, where $\pi(s) = a$ denotes that action a is always taken in state s. The value of a state s under policy π is the expected discounted future reward when starting in state s and executing π thereafter. It satisfies

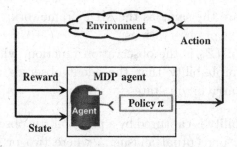

Fig. 4.1. An MDP agent interacting with its environment.

the Bellman equation (Bellman, 1957)

$$V^\pi(s) = R(s, \pi(s)) + \gamma \sum_{s' \in S} T(s' \mid s, \pi(s)) V^\pi(s') \qquad (4.1)$$

An MDP has at least one optimal policy, denoted π^*, which maximizes the value, i.e., $V^*(s) = V^{\pi^*}(s) \geq V^\pi(s)$, for all states $s \in S$. The optimal state value function V^*, the value function associated with any optimal policy π^*, satisfies the Bellman optimality equation

$$V^* = H_{\text{MDP}} V^* \qquad (4.2)$$

where $H_{\text{MDP}} V^*(s) = \max_{a \in A} Q^*(s, a)$ is the Bellman backup operator for MDPs, with the optimal state–action value function

$$Q^*(s, a) = R(s, a) + \gamma \sum_{s' \in S} T(s' \mid s, a) V^*(s') \qquad (4.3)$$

Thus, we can use $\pi^*(s) \leftarrow \arg\max_{a \in A} Q^*(s, a)$ to select an optimal action given the optimal Q-function $Q^*(s, a)$.

4.2.2 *POMDPs*

POMDPs generalize MDPs to problems where the agent cannot completely observe the underlying state due to noisy or incomplete sensor information (Kaelbling *et al.*, 1998). It can be defined as a tuple $(S, A, T, R, Z, \Omega, \gamma)$:

- S, A, T, R, γ are the same as the MDP framework;
- Z is a finite set of observations;
- $\Omega : S \times A \to \Pi(Z)$ is the observation function, where $\Omega(z \mid a, s')$ denotes the probability that the agent observes z when taking action a and arriving in state s'.

Partial observability is captured by a probabilistic observation model. It can lead to "perceptual aliasing," where two or more identical observations require different actions (Chrisman, 1992).

4.2.2.1 *Belief updating*

Because the underlying state is not fully observable, the optimal policy may not be Markovian with respect to the observations. In other words, a direct mapping of observations to actions is not sufficient for optimal behavior. The agent must consider the complete history of the past actions and observations to select a desirable action. The explicit representation of the past can require an arbitrary amount of memory, but it can be avoided. It is possible to summarize all relevant information from previous actions and observations in a discrete probability distribution over the state space S, called a belief state (or belief). A belief state b_t gives the probability that the state is s at time t

$$b_t(s) = P(s_t = s \mid z_t, a_{t-1}, z_{t-1}, \ldots, a_0, b_0) \qquad (4.4)$$

where the initial belief b_0 provides a probability distribution over initial states at time $t = 0$. A belief b satisfies $b(s) \in [0, 1]$ for every $s \in S$ and $\sum_{s \in S} b(s) = 1$. A belief state space $\mathcal{B}$, also called the belief simplex, is the set of all possible beliefs. When the agent takes action a at belief b and receives observation z, it will arrive at a new belief $b^{a,z}$:

$$b^{a,z}(s') = \frac{1}{P(z \mid b, a)} \Omega(z \mid a, s') \sum_{s \in S} T(s' \mid a, s) b(s) \qquad (4.5)$$

where $P(z \mid b, a) = \sum_{s' \in S} \Omega(z \mid a, s') \sum_{s \in S} T(s' \mid a, s) b(s)$. The process of computing the new belief is known as belief updating.

4.2.2.2 *Belief MDPs*

A POMDP can be viewed as a belief MDP. The set of states is the belief space $\mathcal{B}$, and the set of actions is exactly the same as for the POMDP. The state-transition function $T(b' \mid b, a) = \sum_{z \in Z} P(b' \mid b, a, z) P(z \mid b, a)$ where $P(b' \mid b, a, z) = \delta_{b'}(b^{a,z})$ with δ being the Kronecker delta function. The reward function is $R(b, a) = \sum_{s \in S} b(s) R(s, a)$.

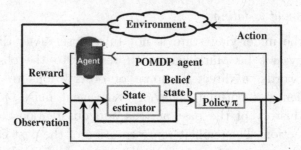

Fig. 4.2. A POMDP agent interacting with its environment.

4.2.2.3 *Policies and value functions*

As shown in Fig. 4.2, a POMDP agent consists of two components. The state estimator component is used to update the belief based on the last action, the current observation, and the previous belief. The policy component is used to generate actions based on the belief. The agent's goal remains to find an optimal policy π^* that maximizes the expected discounted future reward.

A POMDP policy π induces a value function $V^\pi(b)$ that specifies the expected discounted future reward of executing π starting from b

$$V^\pi(b) = E_\pi \left[\sum_{t=0}^{\infty} \gamma^t R(b_t, \pi(b_t)) \mid b_0 = b \right] \qquad (4.6)$$

Like MDPs, the optimal value function V^* in POMDPs also satisfies the Bellman optimality equation

$$V^* = H_{\text{POMDP}} V^* \qquad (4.7)$$

where $H_{\text{POMDP}} V^*(b) = \max_{a \in A} Q^*(b, a)$ is the Bellman backup operator for POMDPs, with $Q^*(b, a) = R(b, a) + \gamma \sum_{z \in Z} P(z \mid b, a) V^*(b^{a,z})$. Thus, the optimal action can be extracted by $\pi^*(b) \leftarrow \arg\max_{a \in A} Q^*(b, a)$.

While the belief space $\mathcal{B}$ is continuous, the value function for a POMDP in both the finite and infinite horizon cases can be approximated arbitrarily closely as the upper envelope of a

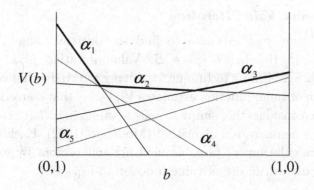

Fig. 4.3. Value function representation of a POMDP with two states. The value function $V(b)$ is indicated by the bold black line. The vectors α_4 and α_5 are totally dominated by other vectors and can be ignored in constructing $V(b)$.

finite number of $|S|$-dimensional hyper-planes $\Gamma = \{\alpha_1, \alpha_2, \ldots, \alpha_n\}$ (Sondik, 1971)

$$V^{\Gamma}(b) = \max_{\alpha \in \Gamma}(\alpha \cdot b) \tag{4.8}$$

where $\alpha \cdot b$ is the standard inner product of a hyperplane α, called an α-vector, and b. Each α-vector is associated with an action. Given a set of α-vectors, the policy can be executed by selecting the action corresponding to the best α-vector at the current belief.

Figure 4.3 shows a value function representation of a POMDP with two states. In such an environment, a belief consists of a vector of two non-negative numbers, $b = (b(s_1), b(s_2))$, satisfying $b(s_1) + b(s_2) = 1$. The vector set Γ consists of five vectors: $\{\alpha_1, \alpha_2, \alpha_3, \alpha_4, \alpha_5\}$. Their upper envelope represents the value function $V(b)$ and is drawn with the bold black line.

4.3   Approaches to Offline Planning

Offline approaches generate a policy over the entire belief space prior to execution. Such approaches can require a significant amount of computation, but during execution they tend to be quite efficient. This section discusses offline approaches that use exact and approximate value iteration.

4.3.1 *Exact Value Iteration*

Exact solution methods aim to find the optimal value $V^*(b)$ for all beliefs in the belief space $\mathcal{B}$. Value iteration is a common dynamic programming technique for solving POMDPs that computes a sequence of value-function estimates $V_1, V_2, \ldots$ that converge to V^*. First, we consider the simplest exact value iteration, called the Monahan's enumeration algorithm (Monahan, 1982). Each iteration involves calculating all possible next-horizon vectors by exploiting the particular structure of value function in Eq. (4.8)

$$H_{\text{POMDP}}\Gamma_n = \bigcup_{a \in A} \Gamma_n^a \tag{4.9}$$

$$\Gamma_n^a = \bigoplus_{z \in Z} \Gamma_n^{a,z} \tag{4.10}$$

$$\Gamma_n^{a,z} = \left\{ \frac{1}{|Z|} R^a + \alpha^{a,z} : \alpha \in \Gamma_n \right\} \tag{4.11}$$

$$\alpha^{a,z}(s) = \sum_{s' \in S} \Omega(z \mid a, s') T(s' \mid s, a) \alpha(s') \tag{4.12}$$

where R^a is a vector representation of the reward function, i.e., $R^a(s) = R(s, a)$, the cross sum of sets is defined as: $\Gamma' \bigoplus \Gamma'' = \{\alpha_1 + \alpha_2 \mid \alpha_1 \in \Gamma', \alpha_2 \in \Gamma''\}$, and the $\alpha^{a,z}$ vectors in Eq. (4.12) resulting from back-projecting $\alpha \in \Gamma_n$ for a particular a and z. As suggested in Eqs. (4.10) and (4.11), we have $|\Gamma_n^{a,z}| = |\Gamma_n|$ and $|\Gamma_n^a| = \prod_z |\Gamma_n^{a,z}| = |\Gamma_n|^{|Z|}$. Thus, there are $|A||\Gamma_n|^{|Z|}$ vectors in $H_{\text{POMDP}}\Gamma_n$, indicating that the set of α-vectors grows exponentially with every iteration. However, many vectors in them do not influence the agent's policy. For example, in Fig. 4.3, α_4 and α_5 are dominated vectors since removing them from the α-vector set does not change the shape of the value function. The algorithm prunes dominated vectors via linear programming, after computing $H_{\text{POMDP}}\Gamma_n$: $\Gamma_{n+1} = \text{Prune}(H_{\text{POMDP}}\Gamma_n)$ (White, 1991).

Finding dominated vectors can be expensive and has long been a key limitation in applying POMDPs to practical domains. Incremental pruning methods (Cassandra, 1998; Cassandra *et al.*,

1997; Feng and Zilberstein, 2004) exploit the fact that

$$\text{Prune}(\Gamma' \oplus \Gamma'' \oplus \Gamma''') = \text{Prune}(\text{Prune}(\Gamma' \oplus \Gamma'') \oplus \Gamma''') \qquad (4.13)$$

to repeatedly prune dominated vectors during the generation procedure, leading to better performance. The basic incremental pruning algorithm computes V_{n+1}^{Γ} in the Γ_{n+1} representation as follows:

$$\Gamma_{n+1} = \text{Prune}\left(\bigcup_{a \in A} \Gamma_n^a\right) \qquad (4.14)$$

$$\Gamma_n^a = \text{Prune}\left(\bigoplus_{z \in Z} \Gamma_n^{a,z}\right) \qquad (4.15)$$

$$= \text{Prune}(\dots \text{Prune}(\text{Prune}(\Gamma_n^{a,1} \oplus \Gamma_n^{a,2}) \oplus \Gamma_n^{a,3})\dots) \qquad (4.16)$$

Other exact value iteration algorithms search in each value-iteration step for a minimal set of beliefs in $\mathcal{B}$ that generates the necessary set of vectors for the next-horizon value function. First, they compute the next-horizon vector α^b in $H_{\text{POMDP}}\Gamma_n$ such that

$$\alpha^b = \operatorname*{arg\,max}_{\alpha \in H_{\text{POMDP}}\Gamma_n} b \cdot \alpha \qquad (4.17)$$

From the given value function V_n^{Γ} in the Γ_n representation, we can derive the vector $\text{Backup}(b, V_n^{\Gamma})$

$$\text{Backup}(b, V_n^{\Gamma}) = \operatorname*{arg\,max}_{\alpha_a^b \,:\, a \in A, \alpha \in \Gamma_n} b \cdot \alpha_a^b \qquad (4.18)$$

$$\alpha_a^b = R^a + \gamma \sum_{z \in Z} \operatorname*{arg\,max}_{\alpha^{a,z} \,:\, \alpha \in \Gamma_n} b \cdot \alpha^{a,z} \qquad (4.19)$$

where $\alpha^{a,z}$ is defined as Eq. (4.12). Second, they employ this point-based Bellman backup operator to compute a complete value function for the next horizon, i.e., locating the (minimal) set of "witness" beliefs such that $\bigcup_b \text{Backup}(b, V_n^{\Gamma}) = \text{Prune}(H_{\text{POMDP}}\Gamma_n)$. Each witness belief testifies the existence of a non-empty region of belief space over which α^b is maximal. A family of exact algorithms (Kaelbling *et al.*, 1998; Littman, 1996) focus on finding such a minimal set of beliefs efficiently.

In general, however, exact solution methods are not sufficient for a POMDP problem with more than a few dozen states, actions, and observations. A major cause of intractability is the aim of computing the optimal action for every possible belief in $\mathcal{B}$. It is possible that in the worst case the number of α-vectors in V_n^Γ grows exponentially in the planning horizon n. In fact, it has been shown that computing an optimal policy for finite-horizon POMDPs are PSPACE-complete (Papadimitriou and Tsitsiklis, 1987), a complexity class that includes NP-complete problems and possibly even more difficult problems. Infinite-horizon POMDPs are undecidable (Madani *et al.*, 1999). As a result, research has focused on efficient methods for computing approximate solutions for much larger problems (Hauskrecht, 2000; Pineau *et al.*, 2006a; Shani *et al.*, 2013; Smith, 2007).

4.3.2 *Approximate Value Iteration*

A significant breakthrough in POMDP research was the introduction of point-based value iteration (PBVI) (Pineau *et al.*, 2003), which can accommodate complex problems with many thousands of states. PBVI bounds the number of α-vectors in its representation of the approximate value function by collecting a subset of beliefs in the reachable belief space (a set of beliefs that can be actually encountered by interacting with the environment). It applies the point-based Bellman backup operator in Eq. (4.18) to the collected belief subset (Algorithm 4.1). Some important variants of PBVI compute lower and upper bounds on the optimal value function V^* and use them in the belief collection stage. This section introduces heuristic methods for initializing the lower and upper bounds of V^*, and then discusses the original PBVI approach and variants.

Algorithm 4.1 Generic point-based approach

1: **while** Planning termination not met **do**
2:      Collect B, a subset of the reachable belief space
3:      Update approximate value function over B

4.3.2.1 *Methods for initializing the value function*

The blind policy method (Hauskrecht, 2000; Smith and Simmons, 2005) is often used to initialize the lower bound of V^*. Here, "blind" means the agent always executes the same action. In a POMDP problem, there are $|A|$ blind policies, and the value function of each blind policy can be represented as an α-vector. The upper envelope of these α-vectors forms a lower bound of V^*. Each α-vector α^a in the lower bound can be computed iteratively

$$\alpha_{t+1}^a(s) = R(s,a) + \gamma \sum_{s' \in S} T(s' \mid s,a)\alpha_t^a(s') \qquad (4.20)$$

where $\alpha_0^a(s) = \min_{s \in S} \frac{R(s,a)}{1-\gamma}$ for all $s \in S$.

MDP, QMDP, and fast informed bound (FIB) are three methods for computing an upper bound of V^* using the underlying MDP. The MDP method obtains the optimal value function of the underlying MDP of a POMDP in an iterative way

$$\alpha_{t+1}(s) = \max_{a \in A} \left[R(s,a) + \gamma \sum_{s' \in S} T(s' \mid s,a)\alpha_t(s') \right] \qquad (4.21)$$

where $\alpha_0(s) = 0$ for all $s \in S$. Since the underlying MDP does not take into account partial observability, its optimal value function is obviously an upper bound of V^* in the corresponding POMDP. The QMDP method (Littman *et al.*, 1995) is a slight variation of the MDP method. It assumes that the problem becomes fully observable after a single step

$$\alpha_{t+1}^a(s) = R(s,a) + \gamma \sum_{s' \in S} T(s' \mid s,a) \max_{a' \in A} \alpha_t^{a'}(s') \qquad (4.22)$$

where $\alpha_0^a(s) = 0$ for all $s \in S$ and $a \in A$. The FIB method takes into account, to some extent, the partial observability of the environment. Its α-vector update process is

$$\alpha_{t+1}^a(s) = R(s,a) + \gamma \sum_{z \in Z} \max_{a' \in A} \sum_{s' \in S} \Omega(z \mid a,s')T(s' \mid s,a)\alpha_t^{a'}(s')$$

$$(4.23)$$

where α_0^a can be initialized to the α-vector found by QMDP at convergence. FIB provides a tighter upper bound than the QMDP method.

4.3.3 *Point-based Value Iteration Methods*

Two key ideas in point-based methods are to (1) compute approximate solutions by sampling only a limited subset of the reachable belief space $\mathcal{R}(b_0)$, denoted B, and (2) update not only the value but also its gradient (the α-vector) at each sampled belief $b \in B$. The intuition is that in many problems $\mathcal{R}(b_0)$ forms a low dimensional manifold in $\mathcal{B}$, and thus can be covered densely enough by a relatively small number of beliefs. The α-vector representation of the value function can guarantee the resulting policy generalizes well and can be effective for beliefs outside the set B.

As mentioned before, the reachable belief space $\mathcal{R}(b_0)$ is a set of beliefs that are reachable from the initial belief b_0 under any sequence of actions and (non-zero probability) observations. It can be represented as an AND/OR belief tree $T_{\mathcal{R}}$ rooted at b_0 (see Fig. 4.4). The nodes of $T_{\mathcal{R}}$ correspond to beliefs in $\mathcal{R}(b_0)$. The edges correspond to action–observation pairs. Thus, the new belief $b^{a,z}$ obtained by Eq. (4.5) is connected to its parent b by an edge (a, z) in $T_{\mathcal{R}}$. Further, we can define the set of beliefs that are reachable from b_0 by following optimal policies π^* as $\mathcal{R}^*(b_0)$. Since all optimal policies are only a small part of all possible policies, $\mathcal{R}^*(b_0)$ is a small subset of $\mathcal{R}(b_0)$.

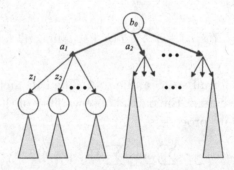

Fig. 4.4. The belief tree $T_{\mathcal{R}}$ rooted at the initial belief b_0.

Usually, $\mathcal{R}^*(b_0)$ is much smaller than $\mathcal{R}(b_0)$ and $\mathcal{R}(b_0)$ is much smaller than $\mathcal{B}$.

The original point-based value iteration method, known as PBVI (Pineau *et al.*, 2003), uses an approximate backup operator $\widetilde{H}_{\text{PBVI}}\Gamma$ instead of $H_{\text{POMDP}}\Gamma$. With each value backup, it computes

$$\widetilde{H}_{\text{PBVI}}\Gamma = \bigcup_{b \in B} \text{Backup}(b, V^{\Gamma}) \tag{4.24}$$

using a fixed set B sampled in $\mathcal{R}(b_0)$. PBVI begins with a small set of initial beliefs B_0 (i.e., $B_0 = \{b_0\}$), performs a number of backup stages, as shown in Eq. (4.24), on B_0, expands B_0 to B_1 by sampling more beliefs, performs backups again, and repeats this process until a stopping criterion is reached. The set B_{t+1} is expanded by simulating actions for every $b \in B_t$, maintaining only the new beliefs that are the most distant from all other beliefs already in B_{t+1}. This sampling scheme is a locally greedy selection strategy that tries to make B_t spread as evenly as possible across $\mathcal{R}(b_0)$.

The original PBVI algorithm focuses on carefully expanding B because B governs the size of the value function and the expansion requires significant computation time. In contrast with PBVI, PERSEUS collects a large B using a random exploration strategy, but only uses a randomly selected subset of B, denoted $\widetilde{B}$ to perform Bellman backup operators at each iteration

$$\widetilde{H}_{\text{PERSEUS}}\Gamma = \bigcup_{b \in \widetilde{B}} \text{Backup}(b, V^{\Gamma}) \tag{4.25}$$

ensuring that $V^{\Gamma}(b') \leq \widetilde{H}_{\text{PERSEUS}}V^{\Gamma}(b'), \quad \forall b' \in B \tag{4.26}$

Thus, $\widetilde{H}_{\text{PERSEUS}}$ is a randomized approximate backup operator that increases (or at least does not decrease) the values over B. Such an operator results in compact value functions, allowing for the use of much larger B, which has the potential to cover $\mathcal{R}(b_0)$ more densely.

The random belief gathering and the random order of backups over the collected beliefs in PERSEUS may be problematic in many complex domains. An alternative is to use a trial-based value iteration algorithm, which gathers promising beliefs in $\mathcal{R}(b_0)$ in a trial-based way, maintaining the visited order of beliefs through the trial, and backing them up in reverse order. This approach allows some beliefs to be updated more than others, based on the assumption that the value function accuracy around the initial belief b_0 is more important than other beliefs in $\mathcal{R}(b_0)$ (Shani *et al.*, 2007; Smith, 2007). Trial-based value iteration algorithms execute a series of trials until a stopping criterion is satisfied. During each trial, a trajectory starting from b_0 is created using a heuristic, and only the beliefs in the trajectory are backed up. This heuristic can help identify important beliefs.

Heuristic search value iteration (HSVI) (Smith and Simmons, 2004, 2005) maintains both lower and upper bounds over V^* and Q^*, denoted V^L, V^U, Q^L, and Q^U, respectively, and uses them to decide which child of the current belief to visit as it explores forward from b_0. Starting from parent belief b, HSVI must choose an action a^* and an observation z^* to arrive in a child belief. It always chooses the action with the highest upper bound

$$a^* \leftarrow \arg\max_{a \in A} Q^U(b, a) \qquad (4.27)$$

Such an action selection strategy helps ensure convergence because the upper bound $Q^U(b, a^*)$ will drop below the upper bound of another action if the action a^* is suboptimal. HSVI uses the excess uncertainty at a belief b with depth d_b in $T_\mathcal{R}$ (i.e., the number of actions from b_0 to b), denoted $\text{excess}(b, d_b) = V^U(b) - V^L(b) - \frac{\epsilon}{\gamma^{d_b}}$, for observation selection

$$z^* \leftarrow \arg\max_{z \in Z} \left[P(z \mid b, a^*) \cdot \text{excess}(b^{a^*, z}, d_b + 1) \right] \qquad (4.28)$$

Such an observation selection strategy can gather beliefs with the greatest contribution to the excess uncertainty at its parent beliefs.

HSVI uses the usual vector set representation for the lower bound and a belief-value mapping representation for the upper bound. Each mapping consists of a belief b and the projection of b onto the convex hull of the upper bound. The reason of not using a vector set in representing the upper bound is that updating by adding a vector has no effect of bringing the bound closer to V^*. Updating of the upper bound requires solving linear programs and generally consumes significant computation time. To simplify solving linear programs, approximate projection techniques, i.e., the sawtooth approximation (Hauskrecht, 2000; Smith and Simmons, 2005), are often considered to speed up upper bound updates. Sawtooth treats the value function as a set of down-facing pyramids. HSVI periodically prunes dominated elements in both the lower bound vector set and the upper bound belief/value set.

In HSVI, the termination condition of forward exploration during a trial is $excess(b, d_b) \leq 0$. HSVI guarantees that we find an approximately optimal policy starting from b_0 after a finite number of iterations, although this number is exponential in the maximum length of a trial.

With the insight that the key difficulty of the computation lies in computing a proper cover[1] of $\mathcal{R}^*(b_0)$ (Hsu *et al.*, 2007; Kurniawati *et al.*, 2008) built the SARSOP algorithm, which stands for successive approximations of the reachable space under optimal policies. In contrast with HSVI, SARSOP uses a modified termination condition to allow selective deep sampling. That is, it uses a simple learning technique to predict the optimal value $V^*(b)$ for some beliefs with $excess(b, d_b) \leq 0$, and continues down a sampling path beyond the belief if it predicts that doing so is likely to lead to an improvement in the lower bound at the initial belief b_0. In comparison with the α-vector pruning strategy in HSVI, the pruning step in SARSOP is more aggressive, helping to improve the computational efficiency of point-based backups. The so-called δ-dominance pruning strategy tries to prune an α-vector if it is dominated by other α-vectors over $\mathcal{R}^*(b_0)$, rather than the entire belief space $\mathcal{B}$. Using these more effective sampling and pruning strategies, SARSOP yields better performance than HSVI and its variants on several benchmark problems.

Packing-guided value iteration (PGVI) (Zhang *et al.*, 2014) can be considered as an extension of SARSOP. Several aspects are the same, including action selection, trial-based forward exploration

---

[1]See Definition 4.1 in Section 4.5.

in a trial-based way, and the procedure of point-based backups at sampled beliefs. Similar to SARSOP, PGVI is also a practical algorithm developed with insights from the theoretical analysis of the complexity of approximate POMDP planning. One key idea behind the theoretical analysis of PGVI is to build a separate packing[2] of sampled beliefs at each depth of the search tree. PGVI used the packing set to alleviate the curse of history in both theoretical and practical aspects. First, it controls the number of sampled beliefs at each depth of the search tree so that it does not grow exponentially in the depth of the tree. This implies that PGVI converges with a number of point-based backups that is quadratic rather than exponential in the planning horizon h. Second, it can be used to identify interesting parts of the reachable belief space that are sparsely packed and to sample new beliefs and perform point-based backups there. PGVI achieves the second one by modifying the observation selection in SARSOP to enable better exploration. Specifically, it prefers to sample beliefs at depth i which are far away from the beliefs in the packing of sampled beliefs at depth i and that no belief in its δ-region has performed point-based backup recently.

In the forward exploration phase of the above trial-based algorithms, such as HSVI, SARSOP, and PGVI, at belief b, only the child belief reachable through the observation that has the highest potential impact is explored. Zhang *et al.* (2015) provided a new method, called Palm LEAf SEarch (PLEASE), allowing for the adaptive selection of multiple observations when their heuristic values are close to the highest one. PLEASE uses the heuristics of beliefs to select promising beliefs in deep levels during the palm leaf search process. Compared with existing trial-based algorithms, PLEASE can reduce the time required for propagating the bound improvements of beliefs deep in the search tree to the root because of fewer point-based value backups. In general, PLEASE is a good

---

[2]Formally, a δ-packing of a set B in a metric space is a set of points P in B such that for any two points $p_1, p_2 \in P$, $\|p_1 - p_2\| > \delta$.

choice for problems with large observation spaces and that require searching deeply to find a near-optimal solutions.

4.4 Approaches to Online Planning

The offline approaches discussed in the previous section use the point-based backup operator to update an approximate value function and use the function to generate a policy defining which action to execute in the whole space $\mathcal{B}$. The use of point-based backups makes the resulting policy generalize well in $\mathcal{B}$, but with significant computational cost. An online approach does not use the expensive backup operator. To return a good local policy from the current belief, it only computes values at beliefs that are reachable from the current belief. Consequently, the overall planning time for online approaches is normally less compared to offline algorithms.

Algorithm 4.2 is a general framework for online POMDP planning algorithms. It can be divided into a planning phase (lines 4–7) and an execution phase (lines 8–10). The two phases are interleaved at each time step. In the planning phase, the approach proceeds by first selecting the next fringe belief to pursue the forward search via the ChooseNextNodeToExpand function. The Expand subroutine constructs the next reachable beliefs under the selected fringe belief and evaluates the approximate values for them. Finally, the UpdateAncestors function is used to propagate the new approximate value of the expanded belief to its

---

Algorithm 4.2 Generic online approach

1: $b_c \leftarrow b_0$;
2: Build an AND/OR tree to contain b_c at the root
3: **while** Execution termination condition not met **do**
4:     **while** Planning termination not met **do**
5:         $b^* \leftarrow$ ChooseNextNodeToExpand()
6:         Expand(b^*)
7:         UpdateAncestors(b^*)
8:     Take best action a for b_c and receive an observation z
9:     $b_c \leftarrow b_c^{a,z}$
10:    Update the tree so that b_c is the new root

---

ancestors. We briefly overview three categories of online approaches: branch-and-bound pruning, heuristic search, and Monte Carlo tree search. These approaches are different only in the implementation of the ChooseNextNodeToExpand and Expand subroutines.

4.4.1 *Branch and Bound*

The branch-and-bound pruning is a general search technique that uses knowledge of the lower and upper bounds on the optimal value function to prune portions of the search tree that are known to be suboptimal. Specifically, for two actions a_1 and a_2 in belief b, the inequality $Q^U(b, a_1) \leq Q^L(b, a_2)$ implies that $Q^*(b, a_1) \leq Q^*(b, a_2)$. Thus, the action a_1 is suboptimal in b and no belief reached by taking action a_1 in b will be considered.

Real-time belief space search (RTBSS) (Poupart, 2005) is an online algorithm that uses the branch-and-bound pruning technique. It computes the lower and upper bounds offline, and RTBSS's efficiency depends largely on their precision. Its ChooseNextNodeTo-Expand simply returns the current belief b_c, and its UpdateAncestors does not need to perform any operation because b_c has no ancestor. The Expand subroutine expands the AND/OR tree in a depth-first fashion, up to some predetermined search depth D, using branch-and-bound pruning. It expands the actions in descending order of their upper bound $Q^U(b, a)$ and stops the expansion of all remaining actions once an action that can be pruned is found.

4.4.2 *Heuristic Search*

Like RTBSS, online heuristic search algorithms also maintain both lower and upper bounds on the value of each belief node in the tree. Instead of using these bounds in branch-and-bound pruning, heuristic search algorithms use them to select the best fringe beliefs to expand via the ChooseNextNodeToExpand subroutine. The Expand subroutine in these algorithms performs a one-step lookahead under the selected fringe node.

There are several online heuristic search algorithms in the literature: the Satia and Lave's approach (Satia and Lave, 1973), bounded, incremental POMDP (BI-POMDP) (Washington, 1997), anytime error-minimization search (AEMS) (Ross and Chaib-Draa, 2007), and factored hybrid heuristic online planning (FHHOP) (Zhang and Chen, 2012). One key difference among them is the specific heuristic used to choose the next fringe node to expand in the AND/OR tree. Usually the fringe node with the highest heuristic value is selected to expand. Heuristic values associated with fringe nodes should indicate the importance level of expanding a particular node to improve the solution. Thus, an online search algorithm using a good heuristic can make good decisions by expanding as few belief nodes as possible. The AEMS heuristic encourages exploration of fringe beliefs that have loose bounds and high probability to be encountered by actions with the highest upper bound. FHHOP adopts a hybrid strategy that depends on AEMS's heuristic and a heuristic function constructed from the lower bound.

4.4.3 *Monte Carlo Tree Search*

Unlike branch-and-bound search and heuristic search algorithms, Monte Carlo algorithms (Kocsis and Szepesvári, 2006; Silver and Veness, 2010) do not use lower and upper bounds. They estimate the value of each belief node in the search tree using Monte Carlo simulation. For each simulation, the start state is sampled from the current belief, and sequences of states, observations, and rewards are sampled from a black box simulator $\mathcal{G}$, $(s_{t+1}, z_{t+1}, r_{t+1}) \sim \mathcal{G}(s_t, a_t)$. Thus, in Monte Carlo algorithms, the POMDP dynamics are encapsulated in the black box and cannot be used directly. Monte Carlo algorithms are very appealing in problems that are too large or too complex to represent with explicit probability distributions.

Monte Carlo tree search (MCTS) (Browne *et al.*, 2012; Kochenderfer, 2015; Kocsis and Szepesvári, 2006) is one of the most successful sampling-based online approaches in recent years. It has outperformed previous planning approaches in a wide variety of challenging tasks, including Go (Silver *et al.*, 2016) and general game playing (Finnsson and Björnsson, 2008; Genesereth and Thielscher, 2014). It was proposed for fully observable domains, and then extended to partially observable domains. In the original MCTS, the ChooseNextNodeToExpand function is used to select a promising belief node in the search tree using the upper confidence bound (UCB) heuristic. That is, the action branch that maximizes

$$Q(b, a) + c\sqrt{\frac{\log N(b)}{N(b, a)}} \qquad (4.29)$$

is always searched. Here, $N(b, a)$ is the number of times that action a has been taken from belief b, $N(b) = \sum_{a \in A} N(b, a)$, and c is a parameter that controls the degree of exploration in the search. The second term is an exploration bonus that encourages selecting action branches that have not been explored frequently. Its Expand subroutine expands the search tree from the selected node by adding one node as its child, and then the expanded node is selected as the start of a simulation, or a rollout, of a complete episode where actions are selected by a rollout policy. Typically, rollout policies are stochastic and do not have to be close to optimal. The result of the simulated episode is used to update the value estimate of the expanded node, and the new approximate value is propagated to its ancestors by the UpdateAncestors subroutine.

Partially observable Monte Carlo planning (POMCP) (Silver and Veness, 2010) is an online planning algorithm for POMDPs based on MCTS. It uses Monte Carlo sampling to break the curse of dimensionality and the curse of history by sampling start states from the current belief, and by sampling histories using a black box simulator. It uses UCB for action selection to control the exploration–exploitation tradeoff, but such a heuristic is not always sample efficient. An iterative algorithm called determinized sparse partially observable tree (DESPOT) (Somani *et al.*, 2013) avoids POMCP's worst-case performance by evaluating policies on a small number of sampled scenarios. Its anytime version, the anytime regularized DESPOT (AR-DESPOT) algorithm searches the DESPOT for a policy, while approximately balancing the size of the policy and its estimated value obtained under the sampled scenarios. Dirichlet–Dirichlet–NormalGamma-based POMCP (D^2NP-POMCP) (Bai *et al.*, 2014) is a recently proposed online

MCTS algorithm using Thompson sampling to balance cumulative and simple regrets during online search.

4.5 Covering-Number-based Planning Theories

Finding a meaningful way to characterize the difficulty of approximate POMDP planning is a core theoretical problem in POMDP research. The state-space size and the number of distinct histories are often used as measures for describing the difficulty in POMDP planning in terms of the curses of dimensionality and history, respectively. The success of point-based algorithms tells us that the curse of history plays a more important role in affecting POMDP value iteration convergence than the curse of dimensionality. Thus, it is reasonable to believe the number of possible histories should be a better predictor of the difficulty of solving POMDPs. The notion of the covering number provides a measure of the space of possible histories and has worked as an important driver in the development of point-based algorithms. In this section, we first introduce the covering number concept, then present the main theoretical results that connect the covering number to the complexity of approximate POMDP planning.

4.5.1 *Covering Number*

The mathematical definition of the covering number of a set of points in a metric space is defined as follows:

Definition 4.1. Given a metric space X, a δ-cover of a set $B \subseteq X$ is a set of points $C \subseteq X$ such that for every point $b \in B$, there is a point $c \in C$ with $||b - c|| \leq \delta$. The δ-covering number of B is the size of the smallest δ-cover of B.

Note that the notion of the covering number of the state space appeared previously in (Kakade *et al.*, 2003), where it refers to the number of neighborhoods required for accurate local modeling in large or infinite state MDPs. Hsu *et al.* (2007) extended the concept from MDPs to POMDPs. Intuitively, the covering number of a space B is equal to the minimum number of balls of radius δ needed to cover B. The distance between beliefs is usually measured in the L_1-metric space $\mathcal{B}$: for $b_1, b_2 \in \mathcal{B}$, $||b_1 - b_2||_1 = \sum_{s \in S} |b_1(s) - b_2(s)|$. Since we can cover the whole belief space $\mathcal{B}$ using only a finite number of balls, the covering numbers of search spaces are always finite for all POMDPs with $\delta > 0$.

4.5.2 *Complexity of Approximate Planning*

The covering number is used to measure the sizes of belief spaces. Existing literature established complexity results of approximate planning on the covering numbers of different spaces: the reachable belief space $\mathcal{R}(b_0)$, the optimally reachable belief space $\mathcal{R}^*(b_0)$, and the search spaces reachable under heuristics. We highlight the key results below:

- A near-optimal POMDP solution can be computed in time at most quadratic in the covering number of $\mathcal{R}(b_0)$ (Hsu *et al.*, 2007).

Hence, POMDPs with small covering numbers of $\mathcal{R}(b_0)$ are easy to solve. However, most interesting POMDP applications have very large covering numbers for $\mathcal{R}(b_0)$.

- The problem of finding a near-optimal solution for a POMDP, even with a very small covering number for $\mathcal{R}^*(b_0)$, say polynomial in the state space size, is NP-hard. However, given a proper cover of $\mathcal{R}^*(b_0)$, a near-optimal solution can be computed in time at most quadratic in the cover size (Hsu *et al.*, 2007).

This result suggests that it may be beneficial to use prior knowledge to find a proper cover of $\mathcal{R}^*(b_0)$. In practice, many well-known POMDP planning algorithms such as HSVI and important variants mentioned in Section 4.3 use the lower and upper bounds of the optimal value function in guiding the search towards $\mathcal{R}^*(b_0)$.

- Zhang *et al.* (2014) formalized the search space that is reachable when both upper and lower bound heuristics are available, and showed that a near-optimal solution can be found in time at most quadratic in the covering number for the search space.

This result suggests an avenue for handling practical problems: use domain knowledge to find tight upper and lower bounds that can reduce the covering number of the reachable space under the heuristics.

The proofs of the above theoretical results are based on the Lipschitz condition that the optimal value function V^* satisfies: for any two beliefs, if their distance is small, then their optimal values are also similar. Thus, when the value of a belief b is sufficiently accurate, it can be used to estimate the value of beliefs that are close to b with only small error. By stopping the search when it is near a δ-region that has been searched before, the width of the search tree is bounded by the size of the δ-picking of sampled beliefs, not larger than the $\frac{\delta}{2}$-covering number of $\mathcal{R}(b_0)$ (Hsu *et al.*, 2007).

Additionally, Hsu *et al.* (2007) analyzed the upper bounds of covering numbers of $\mathcal{R}(b_0)$ for several subclasses of POMDPs, for example, POMDPs with fully observable state variables, sparse beliefs, smooth beliefs, and circulant state-transition matrices. More recently, researchers discussed the extensions of the theoretical results on the covering numbers of different search spaces, including the space reachable under inadmissible heuristics, to more general metric spaces (Zhang *et al.*, 2016) and some types of continuous-state POMDPs (Zhang and Liu, 2016).

4.6 Summary

This chapter focuses on POMDPs, a rich mathematical framework for sequential decision making under uncertainty, an essential capability for autonomous agents operating in partially observable stochastic environments. We introduced the basic concept, offline and online planning algorithms, and covering-number based planning theories in the finite POMDP model. The main points are as follows:

- POMDPs are much more difficult to solve exactly than MDPs. They capture the partial observability in an observation function. To act optimally, a belief that summarizes all relevant information from past actions and observations needs to be calculated. A POMDP can be viewed as an MDP in the belief space.

- Exact methods aim to find optimal actions for all beliefs in the belief space. They are inefficient at handling complex problems.
- Approximate methods have the ability to compute successful policies for larger problems. They can be categorized as offline algorithms and online algorithms.
- Offline approaches, such as point-based algorithms, aim to find an optimal policy starting from the initial belief. They devote significant preprocessing time to generate a policy over the whole belief space in the policy-construction phase. They then use the resulting policy to make decisions during the policy-execution phase.
- Point-based methods compute approximate solutions by iteratively sampling a promising belief subset in the reachable belief space and updating the α-vector at each sampled belief.
- Online approaches aim to find an optimal action for the current belief. They do not allocate significant time to preprocessing, but alternate between a time-limited policy-construction step for the current belief and a policy-execution step.
- There are three broad categories of online approaches: branch-and-bound pruning, heuristic search, and Monte Carlo tree search.
- The covering number can be used to measure the sizes of belief spaces. It is a viable measure for characterizing the difficulty of approximate POMDP planning. Recent results have established theoretical connections between the covering number of different spaces and the complexity of approximate POMDP planning.

Chapter 5

Multiagent Reinforcement Learning Algorithms Based on Gradient Ascent Policy

Chengwei Zhang[*], Xiaohong Li[†], Zhiyong Feng[‡]
and Wanli Xue[§]
*School of Computer Science and Technology,
Tianjin University, Tianjin 300350, China*

[*] *chenvy@tju.edu.cn*
[†] *xiaohongli@tju.edu.cn*
[‡] *zyfeng@tju.edu.cn*
[§] *wanlixue@tju.edu.cn*

In this chapter, we will introduce a class of algorithms which are extended from Q-learning (Watkins, 1989) based on gradient ascent policy. The objective of this chapter is to figure out the advantages, the disadvantages, and the scope of application of these algorithms by analyzing their design objectives and mathematical mechanism. This chapter is organized as follows: Section 5.1 gives a brief summary of gradient ascent based Q-learning algorithms, and classifies each algorithm according to their design purpose. In the rest of each section, we introduce and analyze one or two representative algorithms of each classification, respectively: Section 5.2 for the basic gradient ascent Q-learning algorithm, Infinitesimal Gradient Ascent algorithm (IGA) (Singh *et al.*, 2000); Section 5.3 for WoLF-IGA (Bowling and Veloso, 2002) and WPL (Abdallah and Lesser, 2008); and Section 5.4 for the SA-IGA (Li *et al.*, 2016).

5.1 Introduction

In multiagent systems (MAS), learning ability is important for agents to adjust their behaviors adaptively in response to coexisting agents and unknown environments in order to optimize its performance. There are two important elements that play important roles on achieving agents' goals in multiagent learning systems: the reinforcement value updating policy and the action selecting rule. For action selecting rules in multiagent reinforcement learning (MARL), it is important to balance the exploration and the exploitation. The positive feedback in accordance with the estimate updating in MARL and the greedy heuristics help the action selecting policy to find acceptable solution from long term and short term respectively. As an action selecting policy, gradient ascent policy has received extensive investigation in the literature, and many learning strategies based on GA (Abdallah and Lesser, 2008; Bowling, 2004; Bowling and Veloso, 2002; Kapetanakis and Kudenko, 2002b; Li *et al.*, 2016; Singh *et al.*, 2000) have been proposed to facilitate coordination among agents.

The first gradient ascent MARL algorithm is infinitesimal gradient ascent (IGA) (Singh *et al.*, 2000), in which each learner updates its policy toward the gradient direction of its expected payoff. The purpose of IGA is to promote agents to converge to a particular Nash equilibrium in a two-player two-action normal-form game. IGA has proved that agents will converge to Nash equilibrium or if the strategies themselves do not converge, then their average payoffs will nevertheless converge to the average payoffs of Nash equilibrium. Soon after, Zinkevich (2003) proposed an algorithm called generalized infinitesimal gradient ascent (GIGA), which extends IGA to the game with an arbitrary number of actions. Both IGA and GIGA assume that the agent has the knowledge of the (reward) structure of the game, or at least has some mechanism for approximating the gradient of the value function, which is not generally feasible in practice, and can not guarantee the convergence of the algorithm.

All the algorithms can be combined with the win or learn fast (WoLF) heuristic in order to improve performance in stochastic games (Bowling, 2004; Bowling and Veloso, 2002). The intuition behind WoLF principle is that an agent should adapt quickly when it performs worse than expected, whereas it should maintain the current strategy when it receives better payoff than the expected one. By altering the learning rate according to the WoLF principle, a rational algorithm can be made convergent. The shortage of WoLF–IGA or WoLF–GIGA is that these two algorithms require a reference policy, e.g., a policy belonging to an arbitrary Nash equilibrium. The weighted policy learner (WPL) (Abdallah and Lesser, 2008) is another variation of IGA that also modulates the learning rate, but in contrast to IGA–WoLF, it does not require a reference policy. Both the WoLF and WPL are designed to guarantee convergence in stochastic games.

All the aforementioned learning strategies pursue converging to Nash equilibriums under self-play; however, Nash equilibrium solutions may be undesirable in many scenarios. One well-known example is the prisoners dilemma (PD) game, where Nash equilibrium is not the best option for both agents. In such situations, socially optimal outcome is more attractive than Nash equilibrium. Another research line is to design gradient ascent based learning algorithm to converge to socially optimal outcome under self-play. Li *et al.* (2016) propose a gradient ascent-based algorithm (SA–IGA) which augments the basic gradient ascent algorithm by incorporating social awareness into the policy update process. Experimental results show that a SA–IGA agent can achieve higher social welfare than previous algorithms under self-play and is also robust against individually rational opponents.

5.2 Gradient Ascent Algorithms

Gradient ascent is a well-known optimization technique in the field of Machine Learning. Given a well-defined differentiable objective function, the learning process can follow the direction of its gradient in order to find a local optimum. This concept can be adapted for multiagent learning by having the learning agents policies follow the gradient of their individual expected payoff.

5.2.1 *The Original Gradient Ascent Algorithm:*
Infinitesimal Gradient Ascent (IGA)

The first GA–MARL algorithm whose dynamics were analyzed is
the Infinitesimal Gradient Ascent (IGA) (Singh *et al.*, 2000), in
which each learner updates its policy by taking infinitesimal steps
in the direction of the gradient of its expected payoff. A discrete
time algorithm using a finite decreasing evolutionary dynamics of
multiagent learning step size shares these properties. Set π_i and π_{-i}
to be policies of agent i and its opponents. Take $V_i(\pi) : \mathbb{R}^n \to \mathbb{R}$
to be the value function that maps a joint policy $\pi = \langle \pi_i, \pi_{-i} \rangle^T$ to
agent i's expected payoff. The policy update rule for a IGA player i
can be defined as

$$\Delta \pi_i^{t+1} \leftarrow \alpha \frac{\partial V_i(\pi^t)}{\partial \pi_i}$$
$$\pi_i^{t+1} \leftarrow \prod_{[0,1]} \left(\pi_i^t + \Delta \pi_i^{t+1} \right) \tag{5.1}$$

Parameter α is the policy-learning rate and approaches zeros in
the limit $\alpha \to 0$, hence the word Infinitesimal in IGA. Function
$\Pi[0, 1]$ is the projection function mapping the input value to the
valid probability range of $[0, 1]$, used to prevent the gradient moving
the strategy out of the valid probability space. Formally,

$$\prod_{[0,1]} (x) = \arg \min_{z \in [0,1]} |x - z| \tag{5.2}$$

The purpose of IGA is to promote agents to converge to a
particular Nash Equilibrium in a two-player two-action normal-
form game. The generalized infinitesimal gradient ascent (GIGA)
(Zinkevich, 2003), which extends IGA to the game with an arbitrary
number of actions, extends IGA to games with an arbitrary number
of actions. It has been proved that IGA and its generalization GIGA
can converge in games with pure NE. However, both algorithms failed
to converge in games with mixed NE, and therefore may not be
suitable for applications that require a mixed policy.

Combined with Q-learning, researchers propose a practical learning algorithm, which is a simple extension of IGA and is shown

Table 5.1. The policy hill-climbing algorithm (PHC) for player i.

---

(1)   Lets $\alpha, \beta \in (0, 1)$ be learning rates. Initialize,
        $Q_i(a, s)^{\text{a}} \leftarrow 0$, $\pi_i(s, a)^{\text{b}} \leftarrow \frac{1}{|A_i|}$

(2)   Repeat,
        (2.1)   Select action $a \in A_i$ in state s according to mixed strategy $\pi_i(s)$
                with suitable exploration.
        (2.2)   Observing reward r and next state s'. Update Q,

$$Q_i(s, a) \leftarrow (1 - \beta) Q_i(s, a) + \delta \left(r + \beta^{\text{c}} \max_{a'} Q(s', a') \right).$$

        (2.3)   Update π_i according to gradient ascent strategy,

$$\pi_i(s, a) \leftarrow \Pi_{[0,1]} [\pi_i(s, a) - \alpha], \text{ if } a \neq \text{argmax}_{a'} Q(s, a'),$$

$$\pi_i(s, a) \leftarrow 1 - \sum_{a' \neq a} \pi(s, a'), \text{ if } a = \text{argmax}_{a'} Q(s, a').$$

Notes: a: $Q_i(s, a)$ is an estimate of $V_i(a_i)$ in state s, i.e., the expected payoff to the player i if player i plays action a;
b: $\pi_i(s, a)$ is the probability that player i chooses action a in state s, i.e., the policy of player i;
c: γ is the discount factor.

in Table 5.1. The algorithm performs hill-climbing in the space of mixed policies, which is similar to gradient ascent, but does not require as much knowledge. Q-values are maintained just as in normal Q-learning (Watkins, 1989). In addition, the algorithm maintains the current mixed policy. The policy is improved by increasing the probability that it selects the highest valued action according to a learning rate $\alpha \in (0, 1]$.

This technique, like Q-learning, is rational and will converge to an optimal policy if the other players are playing stationary strategies. The algorithm guarantees the Q-values will converge to Q^* with a suitable exploration policy. π will converge to a policy that is greedy according to Q, which is converging to Q^*, and therefore will converge to a best response. Despite the fact that it is rational and has no limit on the number of agents and actions, it still does not show any promise of being convergent.

Improvements of IGA mainly focused on two directions: to improve the convergence properties of IGA, and to improve the social welfare of all agents. On the one hand, the IGA algorithm did not converge in all two-player-two-action games. To improve the convergence properties of IGA, some studies have focused on alterations of the learning rate α of IGA, such as WoLF–IGA (Bowling and Veloso, 2002), PD–WoLF (Banerjee and Peng, 2003) and WPL (Abdallah and Lesser, 2008), while others are trying to improve the expected payoff function $V_i(\pi)$ of IGA, such as IGA-PP (Zhang and Lesser, 2010). On the other hand, Nash equilibrium solutions may be undesirable in many scenarios. From the system designers' perspective, it is desirable if the agents can learn to coordinate toward socially optimal outcomes, while also avoiding being exploited by selfish opponents. To address this issue, some work is being done trying to find the optimal strategy for the group (i.e., social welfare policy) by extending the IGA, such as SA–IGA (Li *et al.*, 2016). Next, we detail the various improved algorithms for IGA.

5.2.2 *Algorithms Improving the Convergence Properties of IGA*

Win or Learn Fast (WoLF) IGA. Algorithm IGA–WoLF (WoLF stands for Win or Learn Fast) (Table 5.2) was proposed (Bowling, 2004; Bowling and Veloso, 2002) in order to improve convergence properties of IGA by using two different learning rates as follows. If a player is getting an average reward lower than the reward it would get for executing its NE strategy, then the learning rate should be large. Otherwise (the player current policy is better than its NE policy), the learning rate should be small. Formally,

$$\Delta \pi_i^{t+1} \leftarrow \alpha_i^t \frac{\partial V_i(\pi^t)}{\partial \pi_i}$$

$$\pi_i^{t+1} \leftarrow \prod_{[0,1]} \left(\pi_i^t + \Delta \pi_i^{t+1} \right) \tag{5.3}$$

$$\alpha_i^t = \begin{cases} \alpha_{\text{lose}} & \text{if } V_i\left(\pi_i^t, \pi_{-i}^t \right) < V_i\left(\pi_i^*, \pi_{-i}^t \right) \\ \alpha_{\text{win}} & \text{otherwise} \end{cases} \tag{5.4}$$

where π_i^* is a reference policy for agent i, e.g., a policy belonging to an arbitrary Nash equilibrium, and $\alpha_{\text{lose}} > \alpha_{\text{win}}$ are the learning rates. The dynamics of IGA–WoLF have been analyzed and proved to converge in all two-player, two-action games (Bowling and Veloso,

Table 5.2. The WoLF policy hill-climbing algorithm (WoLF–PHC) for player i.

---

(1) Lets $\alpha, \delta \in (0, 1)$ be learning rates. Initialize,
$Q_i(a, s) \leftarrow 0$, $\pi_i(s, a) \leftarrow \frac{1}{|A_i|}$, $C_i(s)^{\text{a}} \leftarrow 0$

(2) Repeat,
  (2.1) Same as PHC in Table 5.1.
  (2.2) Same as PHC in Table 5.1.
  (2.3) Update estimate of average policy, $\bar{\pi}^{\text{b}}$,

  $C_i(s) \leftarrow C_i(s) + 1$,

  $\forall a' \in A_i,\ \bar{\pi}_i(s, a') \leftarrow \bar{\pi}_i(s, a') + \frac{1}{C_i(s)}(\pi(s, a') - \bar{\pi}(s, a'))$.

  (2.4) Update π_i according to WoLF gradient ascent strategy. Same as PHC in Table 5.1 (2.3), but with

  $$\alpha_i = \begin{cases} \alpha_{\text{lose}} & \text{if } \sum_{a'} \pi(s, a') Q(s, a')^{\text{c}} < \sum_{a'} \bar{\pi}(s, a') Q(s, a') \\ \alpha_{\text{win}} & \text{otherwise} \end{cases}$$

Notes: a: $C_i(s)$ is the number of times of player i in state s;
b: $\bar{\pi}$ is the cumulative average policy of player i in state s, play the role of the reference policy defined in Eq. (5.4);
c: $\sum_{a'} \pi(s, a') Q(s, a')$ is the estimation of $V_i(\pi_i^t)$. Similar for $\sum_{a'} \bar{\pi}(s, a') Q(s, a')$.

2002). The main limitation of IGA–WoLF is that it assumed each agent knows the NE policy (needed to switch between the two modes of the learning rate).

Policy Dynamics based WoLF (PD–WoLF). Because the conditions of WoLF–IGA in Eq. (5.4) cannot be accurately computed without knowing π_i^*, Banerjee and Peng (2003) propose an alternative criterion of WoLF–IGA, named Policy Dynamics based WoLF (PD–WoLF) (Table 5.3), that can be easily and accurately computed and guarantees convergence. The update function of learning rate α_i^t in PD–WoLF is defined as follows:

$$\alpha_i^t = \begin{cases} \alpha_{\text{win}} & \text{if } \delta^t \delta^{2t} < 0 \\ \alpha_{\text{lose}} & \text{otherwise} \end{cases} \tag{5.5}$$

Valuables δ^t and δ^{2t} are the first- and second-order differences of policy π_i. Formally,

$$\delta^t = \pi^t - \pi^{t-1}$$
$$\delta^{2t} = \delta^t - \delta^{t-1} \tag{5.6}$$

Table 5.3. The PD–WoLF hill-climbing algorithm (PDWoLF–PHC) for player i.

---

(1)   Lets $\alpha, \delta \in (0,1)$ be learning rates. Initialize,
      $Q_i(a,s) \leftarrow 0$, $\pi_i(s,a) \leftarrow \frac{1}{|A_i|}$, $\delta_i(s,a) \leftarrow 0$, $\delta_i^2(s,a) \leftarrow 0$

(2)   Repeat,

      (2.1)   Same as PHC in Table 5.1.

      (2.2)   Same as PHC in Table 5.1.

      (2.3)   Compute

      $$\pi_i^{\text{temp}}(s,a)^{\text{a}} \leftarrow \pi_i(s,a)$$

      (2.4)   Update π_i according to PD–WoLF gradient ascent strategy. Same as PHC in Table 5.1 (2.3), but with

      $$\alpha_i = \begin{cases} \alpha_{\text{win}} & \text{if } \delta_i(s,a)\,\delta_i^2(s,a) < 0 \\ \alpha_{\text{lose}} & \text{otherwise} \end{cases}$$

      (2.5)   Update δ and δ^2,

      $$\delta_i^{\text{temp}}(s,a)^{\text{b}} \leftarrow \pi_i(s,a) - \pi_i^{\text{temp}}(s,a)$$
      $$\delta_i^2(s,a) \leftarrow \delta_i^{\text{temp}}(s,a) - \delta_i(s,a)$$
      $$\delta_i(s,a) \leftarrow \delta_i^{\text{temp}}(s,a)$$

Notes: a: $\pi_i^{\text{temp}}(s,a)$ record the value of $\pi_i(s,a)$ before it was updated;
b: $\delta_i^{\text{temp}}(s,a)$ record the value of $\delta_i(s,a)$ after the update. Both $\pi_i^{\text{temp}}(s,a)$ and $\delta_i^{\text{temp}}(s,a)$ are used to assist in the update of δ and δ^2.

Banerjee and Peng (2003) proved experimentally that PDWoLF–IGA, as an accurately computable version of WoLF–IGA, has a better performance than WoLF–IGA in games with higher actions, for example, the convergence rate.

The Weighted Policy Learner Algorithm (WPL). The weighted policy learner (WPL) (Table 5.4) (Abdallah and Lesser, 2008) is another variation of IGA that also modulates the learning rate, but in contrast to IGA–WoLF it does not require a reference policy. The policy update rule for a WPL player i is defined the same as IGA–WoLF (5.3), except for the learning rate α_i^t,

$$\alpha_i^t = \begin{cases} \alpha \pi_i & \text{if } \dfrac{\partial V(\pi^t)}{\partial \pi_i} < 0 \\ \alpha(1-\pi_i) & \text{otherwise} \end{cases} \tag{5.7}$$

Table 5.4. The Weighted Policy Learner Algorithm (WPL) for player i.

---

(1)   Lets $\alpha, \delta \in (0,1)$ be learning rates. Initialize,
$$Q_i(a,s) \leftarrow 0, \; \pi_i(s,a) \leftarrow \tfrac{1}{|A_i|}$$
(2)   Repeat,
      (2.1)   Same as PHC in Table 5.1.
      (2.2)   Same as PHC in Table 5.1.
      (2.3)   Update π_i according to WPL gradient ascent strategy. Same as
             PHC in Table 5.1 (2.3), but with
$$\alpha_i = \begin{cases} \alpha \pi_i(s,a) & \text{if } a \neq \text{argmax}_{a'} \, Q(s,a') \\ \alpha(1 - \pi_i(s,a)) & \text{otherwise} \end{cases}$$

---

where the update is weighted either by π_i or by $1 - \pi_i$ depending on the sign of the gradient. This means that π is driven away from the boundaries of the policy space. Using this adjustment mechanism, WPL can converge in all two-player, two-action games. The side effect is that the algorithm converges slowly in games with pure NE.

All the aforementioned learning strategies pursue to improving the convergence performance of IGA by adjusting the learning rate α. Next, we introduce an alternative algorithm by adjusting the expected payoff function $V_i(\pi)$.

Gradient Ascent with Policy Prediction (IGA–PP). Zhang and Lesser (2010) propose a gradient-based learning algorithm by adjusting the expected payoff function $V_i(\pi)$ of IGA, named Gradient Ascent with Policy Prediction Algorithm (IGA–PP) (Table 5.5). The key idea behind this algorithm is that a player adjusts its strategy in response to forecasted strategies of the other players, instead of their current ones. The algorithm is specifically designed for two-agent, two-action games. Take $V_{-i}(\pi) : \mathbb{R}^n \to \mathbb{R}$ to be the value function that maps a joint policy π to the expected payoff of agent i's opponent's. The strategy update rules are changed to

$$\Delta \pi_i^{t+1} \leftarrow \alpha \frac{\partial V_i(\pi_i^t, \pi_{-i}^*)}{\partial \pi_i}$$
$$\pi_i^{t+1} \leftarrow \prod_{[0,1]} \left(\pi_i^t + \Delta \pi_i^{t+1} \right)$$

$$(5.8)$$

Table 5.5. The Policy Gradient Ascent with approximate policy prediction (PGA–PP) for player i.

---

(1)   Lets $\alpha, \delta \in (0, 1)$ be learning rates.
       Lets $\lambda \in (0, 1)$ be the derivative prediction length. Initialize,
       $Q_i(a, s) \leftarrow 0$, $\pi_i(s, a) \leftarrow \frac{1}{|A_i|}$

(2)   Repeat,
       (2.1) Same as PHC in Table 5.1.
       (2.2) Same as PHC in Table 5.1.
       (2.3) Compute the average reward $V_i(s)$
              $$V_i(s) \leftarrow \sum_{a'} \pi(s, a') Q(s, a')$$
       (2.3) Update π_i according to gradient ascent with policy prediction
              strategy. Same as PHC in Table 5.1 (2.3), but with
              $$\alpha_i = \begin{cases} \alpha \pi_i(s, a) & \text{if } a \neq \text{argmax}_{a'} Q(s, a') \\ \alpha(1 - \pi_i(s, a)) & \text{otherwise} \end{cases}.$$

---

where π^*_{-i} is the forecasted strategy of agent i's opponent. Take λ as the prediction length of the opponents strategy. π^*_{-i} is defined as follows:

$$\pi^*_{-i} = \prod_{[0,1]} \left(\pi^t_{-i} + \lambda \frac{\partial V_{-i}(\pi^t_i, \pi^t_{-i})}{\partial \pi_{-i}} \right) \qquad (5.9)$$

It has been proved that, in two-player, two-action, general-sum matrix games, IGA–PP in self-play or against IGA would lead players' strategies to converge to a Nash equilibrium. Like other MARL algorithms, besides the common assumption, this algorithm also has additional requirements that a player knows the other players' strategy and current strategy gradient (or payoff matrix) so that it can forecast the other players' strategy.

5.2.3 *Algorithms Improving Social Welfare of IGA*

Social Awareness IGA (SA–IGA). All the aforementioned learning strategies pursue converging to Nash equilibriums under self-play. In this subsection, we propose a novel gradient ascent-based algorithm (SA–IGA) (Li *et al.*, 2016) which augments the basic gradient ascent algorithm by incorporating social awareness into the policy update process. In SA–IGA, apart from learning to maximize its individual payoff, an agent is also equipped with the social awareness such that it can (1) reach mutually cooperative solutions faced with another socially aware opponent (self-play) and (2) behave in a purely individually rational manner against a purely rational opponent.

Specifically, for each agent i, SA–IGA distinguish two types of expected payoffs, namely V_i^{idv} and V^{soc}. The payoff $V_i^{\text{idv}}(\pi)$ and $V^{\text{soc}}(\pi)$ represent the individual and social payoff (the average payoff of all agents) that agent i perceives under the joint strategy π, respectively. The payoff $V_i^{\text{idv}}(\pi)$ follows the same definition as IGA, and the payoff $V^{\text{soc}}(\pi)$ is defined as the average of expected payoffs of all agents. Each agent i adopts a social attitude w_i to reflect its socially-aware degree. The social attitude intuitively models an agent's socially friendly degree towards its partner. Specifically, it is used as the weighting factor to adjust the relative importance between V_i^{idv} and V^{soc}, and agent i's overall expected payoff is defined as follows:

$$V_i(\pi) = (1 - w_i) V_i^{\text{idv}}(\pi) + w_i V^{\text{soc}}(\pi) \qquad (5.10)$$

Each agent updates its policy in the direction of maximizing the value of V based on the Gradient Ascent strategy. Therefore, the policy update rule for a SA–IGA player i is defined the same as IGA in Eq. (5.1). Besides, each agent i's socially aware degree in SA–IGA is adaptively adjusted during each interaction, in response to the relative value of V_i^{idv} and V_i^{soc}. The socially-aware degree w_i is updated as follows:

$$\Delta w_i^{t+1} \leftarrow \alpha_w \left(V_i^{\text{idv}} - V^{\text{soc}} \right)$$
$$w_i^{t+1} \leftarrow \prod_{[0,1]} \left(w_i^t + \Delta w_i^{t+1} \right) \qquad (5.11)$$

If player i's own expected payoff V_i^{idv} exceeds the value of V^{soc}, then player i increases its social attitude w_i, (i.e., it becomes more social-friendly because it perceives itself to be earning more than the average). Conversely, if V_i^{idv} is less than V^{soc} t, then the agent tends to care more about its own interest by decreasing the value of w_i. In SA–IGA, each agent needs to know the policy of its opponent and the payoff matrix, which are usually not available before a repeated game starts. Based on the idea of SA–IGA, we relax the above assumptions and propose a practical multiagent learning algorithm called Socially Aware Policy Gradient Ascent (SA–PGA). The overall flow of SA–PGA is shown in Table 5.6. In SA–PGA, each agent only needs to observe the payoffs of both agents by the end of each round.

With above improvements, Li *et al.* (2016) shows experimentally that an SA–PGA agent can achieve higher social welfare than previous algorithms under self-play and also is robust against individually rational opponents.

Table 5.7 lists an overview of the six discussed algorithms: IGA, WoLF–IGA, PDWoLF–IGA, WPL, PP–IGA, and SA–IGA.

Table 5.6.   SA–PGA for player i.

---

(1)   Let α_π, $\alpha_w \in (0,1)$, $\alpha \in (0,1)$ and be learning rates. Initialize,

$Q_i(a,s) \leftarrow 0$, $\pi_i(s,a) \leftarrow \frac{1}{|A_i|}$, $C_i(s)^a \leftarrow 0$

(2)   Repeat,

    (2.1) Same as PHC in Table 5.1.

    (2.2) Observing reward r and the average of payoffs of all agents r^{all},

$$Q_i^{\text{idv}}(s,a) \leftarrow (1-\beta)\, Q_i^{\text{idv}}(s,a) + \delta\Big(r + \beta\max_{a'} Q_i^{\text{idv}}(s',a')\Big),$$

$$Q_i^{\text{all}}(s,a) \leftarrow (1-\beta)\, Q_i^{\text{all}}(s,a) + \delta\Big(r^{\text{all}} + \beta\max_{a'} Q_i^{\text{all}}(s',a')\Big),$$

$$Q_i(s,a) \leftarrow (1-w_i)\, Q_i^{\text{idv}}(s,a) + w_i Q_i^{\text{all}}(s,a).$$

    (2.3) Update π_i according to gradient ascent strategy. Same as PHC in Table 5.1 (2.3).

    (2.4) Update w_i,

$$V_i^{\text{idv}} = \sum_{a \in A_i} \pi_i(a) Q_i^{\text{idv}}(a).$$

$$V_i^{\text{all}} = \sum_{a \in A_i} \pi_i(a) Q_i^{\text{all}}(a).$$

$$w_i \leftarrow \Pi_{[0,1]}[w_i + \alpha_w\left(V_i^{\text{idv}} - V_i^{\text{all}}\right)].$$

Notes: a: $C_i(s)$ is the number of times of player i in state s;

b: $\bar{\pi}$ is the cumulative average policy of player i in state s, play the role of the reference policy defined in Eq. (5.4);

c: $\sum_{a'} \pi(s,a') Q(s,a')$ is the estimation of $V_i\left(\pi_i^t\right)$. Similar for $\sum_{a'} \bar{\pi}(s,a') Q(s,a')$.

Table 5.7. This table shows an overview of the Gradient Ascent Learning algorithms, rewritten for the specific case of two-agent two-action games.

Algorithms[a]	$\Delta\pi$	Converge or not	Convergence target	Information needs
IGA	$\alpha\partial V_\pi$[a]	No guarantee	NE	Opponent's information
WoLF–IGA	$\partial V_\pi \begin{cases} \alpha_{\text{lose}} & \text{if } V_i(\pi) < V_i(\pi^*) \\ \alpha_{\text{win}} & \text{otherwise} \end{cases}$	Guarantee	NE	Need a reference policy
PDWoLF–IGA	$\partial V_\pi \begin{cases} \alpha_{\text{lose}} & \text{if } \delta^t \delta^{2t} < 0 \\ \alpha_{\text{win}} & \text{otherwise} \end{cases}$	Guarantee	NE	No need
WPL	$\alpha\partial V_\pi \begin{cases} \pi & \partial V_\pi < 0 \\ 1-\pi & \text{otherwise} \end{cases}$	Guarantee	NE	No need
PP-IGA	$\alpha\partial V_{\pi=<\pi_i,\pi^*_{-i}>}$[b]	Guarantee	NE	Opponent's information
SA-IGA	$\alpha\partial V_\pi(w,\pi)$[c]	Guarantee with suitable $w(0)$[d]	Social welfare	Average payoff of all agent

Notes: a: ∂V_π stands for the partial derivative of V_i with respect to π_i;

b: $\partial V_{\pi=<\pi_i,\pi^*_{-i}>}$ stands for the partial derivative of $V_i(\pi_i,\pi^*_{-i})$ with respect to π_i, π^*_{-i} is the forecasted strategy of agent i's opponent;

c: $\partial V_\pi(w,\pi)$ is the partial derivative of $V(w,\pi)$ with respect to π_i, where $V(w,\pi)$ stands for the over all payoff weighted by w between V_i^{idv} and V^{soc};

d: $w(0)$ is the initial value of the socially-aware degree w.

5.3   Dynamics of GA–MARL Algorithms

In this section, we derive the dynamical model of algorithms presented in the previous sections in two-player two-action games, and show their similarity and difference.

In a two-player, two-action, general-sum normal-form game, the payoffs for each player $i \in \{1, 2\}$ can be specified by a matrix as follows:

$$R_i = \begin{bmatrix} r_i^{11} & r_i^{12} \\ r_i^{21} & r_i^{22} \end{bmatrix}$$

Each player i simultaneously selects an action from its action set $A_i = \{1, 2\}$, and the payoff of each player is determined by their joint actions. For example, if player 1 selects the pure strategy of action 1 while player 2 selects the pure strategy of action 2, then player 1 receives a payoff of r_1^{12} and player 2 receives the payoff of r_2^{21}.

Apart from pure strategies, each player can also employ a mixed strategy to make decisions. A mixed strategy can be represented as a probability distribution over the action set and a pure strategy is a special case of mixed strategies. Let $\pi_1 \in [0, 1]$ and $\pi_2 \in [0, 1]$ denote the probability of choosing action 1 by player 1 and player 2, respectively. Given a joint mixed strategy $\pi = (\pi_1, \pi_2)$, the expected payoffs of player $i \in \{1, 2\}$ can be specified as follows:

$$V_i(\pi) = r_i^{11} \pi_i \pi_{-i} + r_i^{12} \pi_i (1 - \pi_{-i}) + r_i^{21} (1 - \pi_i) \pi_{-i}$$
$$+ r_i^{22} (1 - \pi_i)(1 - \pi_{-i}) \tag{5.12}$$

For ease of exposition, we concentrate on unconstrained update equations by removing the policy projection function which does not affect our qualitative analytical results. Any trajectory with linear (nonlinear) characteristic without constraints is still linear (nonlinear) when a boundary is enforced. Combined with definitions in Section 8.2, IGAs update Equation (5.1), which can be simplified to be

$$\pi_i^{t+1} = \pi_i^t + \alpha \frac{\partial V_i(\pi^t)}{\partial \pi_i} = \pi_i^t + \alpha \left(u_i \pi_{-i}^t + c_i \right) \tag{5.13}$$

where u_i and c_i are game-dependent constants having the following values:

$$u_i = r_i^{11} - r_i^{12} - r_i^{21} + r_i^{22}$$
$$c_i = r_i^{12} - r_i^{22}$$

As $\alpha \to 0$, it is straightforward to show that Eq. (5.13) becomes differential. Thus the unconstrained dynamics of the strategy pair as a function of time is simplified by the following system of differential equations:

$$\dot{\pi} = \alpha \left(u_i \pi_{-i} + c_i \right) \tag{5.14}$$

By qualitative analysis of the differential equations (5.14), we can obtain the properties of the algorithm IGA, such as the dynamic trajectory of each agents, the convergence speed, and the position of the equilibrium point. The extension of the dynamics of IGA to other algorithms in the previous section are straightforward.

Table 5.8 lists the dynamics of the six discussed algorithms: IGA, WoLF–IGA, PDWoLF–IGA, WPL, PP–IGA, PP–IGA, and

Table 5.8. The dynamic of the Gradient Ascent Learning algorithms, rewritten for the specific case of two-agent two-action games.

Algorithms	$\dot{\pi}_i$
IGA	$\alpha \left(u_i \pi_{-i} + c_i \right)$
WoLF–IGA	$\left(u_i \pi_{-i} + c_i \right) \begin{cases} \alpha_{\text{lose}} & \text{if } V_i \left(\pi \right) < V_i \left(\pi^* \right) \\ \alpha_{\text{win}} & \text{otherwise} \end{cases}$
PDWoLF–IGA	$\left(u_i \pi_{-i} + c_i \right) \begin{cases} \alpha_{\text{lose}} & \text{if } \delta^t \delta^{2t} < 0 \\ \alpha_{\text{win}} & \text{otherwise} \end{cases}$
WPL	$\alpha \left(u_i \pi_{-i} + c_i \right) \begin{cases} \pi_i & \partial V_\pi < 0 \\ 1 - \pi_i & \text{otherwise} \end{cases}$
PP–IGA	$\alpha \left(u_i \pi_{-i}^* + c_i \right)^{\text{a}}$
SA–IGA	$\alpha \left(u_i + \frac{u_i - u_{-i}}{2} w_i \right) \pi_{-i} + \frac{c_{-i} - c_i}{2} w_i + c_i^{\text{b}}$

Notes: a: π_{-i}^* is the forecasted strategy of agent i's opponent, with $\pi_{-i}^* = \left(1 + \gamma u_{-i} \right) \pi_{-i} + \gamma c_{-i}$;
b: $\dot{w}_i = \beta \left[V_i \left(\pi \right) - V_{-i} \left(\pi \right) \right]$ is the dynamic of i's social attitude w_i, β is the learning rate.

SA–IGA. It is immediately clear from this table that the first four algorithms share the same basic term in their dynamics: the gradient $(u_i \pi_{-i} + c_i)$, while the last two algorithms using a slightly modified gradient function. Depending on the algorithm, the gradient is scaled with a learning speed modulation.

This analysis shows the merits of the evolutionary game theoretic approach to the study of multiagent learning. By deriving mathematical models of the infinitesimal time limit of various learning algorithms, we can formally establish their underlying differences and commonalities.

5.4 Conclusions

In this chapter, we introduced a kind of learning method, gradient ascent Q-learning, and analyzed their dynamic behavior. We divided those methods into methods for achieving Nash equilibrium and methods for achieving social welfare, according to their learning goal. All of those methods are theoretically analyzed by expressing their learning strategy into a set of linear or nonlinear differential equations.

The methods presented in this chapter are widely used reinforcement learning methods. This is probably due to their simplicity: they can be applied with a minimal amount of computation, to experience generated from interaction with an environment; they can be expressed nearly completely by single equations that can be implemented with small computer programs.

Chapter 6

Task Allocation in Multiagent Systems: A Survey of Some Interesting Aspects

Jun Wu[*], Lei Zhang[†], Yu Qiao[‡] and Chongjun Wang[§]

State Key Lab for Novel Software Technology,
Nanjing University, Nanjing Shi, Jiangsu Sheng 210008, China

[*]*wujun@nju.edu.cn*
[†]*zhangl@nju.edu.cn*
[‡]*yuqiao@smail.nju.edu.cn*
[§]*chjwang@nju.edu.cn*

Task allocation is an approach for coordination and cooperation in multiagent systems (MAS). It is an important problem in MAS research and is widely applied in fields such as multirobot systems, crowdsourcing, wireless sensor networks, web services, cloud computing, distributed intrusion detection, and so on. Intrinsically, task allocation is a class of optimization problems, that is, how to allocate a set of tasks to a MAS, in order to achieve a certain objective subject to some constraints. In different applications, the above problem can be formalized as different forms and requires different techniques for designing algorithms. The problem space of task allocation is rather large. However, due to the fact that task allocation is more or less independently studied in several sub-domains of computer science and artificial intelligence, a global view on the problem space of task allocation is still missing. In this chapter, we briefly summarize several important directions of task allocation research, including taxonomy study; allocating tasks that are constrained, complex, or dynamic; allocating tasks to agents that are rational, dynamic, or connected by network; and control models which are currently a spectrum between centralized and distributed models. By the above work, we aim to correlate current task allocation research, preliminarily depict the problem space of task allocation, and locate problems that are well-studied or potentially interesting for future study.

6.1 Introduction

Due to its generality and theoretical importance, in the past decades task allocation was extensively studied as a fundamental problem across a lot of domains such as multirobot systems, wireless sensor networks, crowdsourcing systems, cloud computing systems, social networks, and so on. Typically these work focus on formulating *ad hoc* approaches to solve task allocation problems in a specific domain (or a specific class of systems). It is interesting and of theoretical importance to correlate the related studies and make a general taxonomy on them. In this chapter, we uniformly view such systems as instances of multiagent system and summarize these large amount of research work in the framework of *task allocation in multiagent systems* (MAS). It makes sense since all of such systems are composed of distributed components that are more or less autonomous or self-interested, and some task allocation problems, although studied in different systems, obviously reduce to an identical mathematical problem.

In general, an application in a MAS can be divided into a number of tasks and be executed by different agents; the performance of an application in a MAS is dependent on the allocation of the tasks comprising the application onto the available agents, referred to as the *task allocation problem* (Jiang, 2015). So, basically task allocation is the *class of optimization problems* of determining which agents should execute which tasks in order to achieve the overall system goals. Because of the generality in definition, task allocation problems has a lot of variants different from each other, at least in the following five aspects:

(1) *Properties of the MAS*: The MAS can be instantiated as different application systems, such as multirobot systems, crowdsourcing systems, social networks, etc; each agent can be a rational player

in the sense of game theory,[1] or be a cooperative problem solver that is designed to be truthful; the information of each agent, such as the available resources (including data, processors, computational capacity, disk storage, replicas of popular web objects) and cost of executing a task is public or privately held by the agent; the capacity of an agents is certain or some parameters remain under uncertainty; the agents are fixed or arriving and departing over time; cooperating is available for every subset of agents or constrained by some underlying networks.

(2) *Properties of the task set*: Each task can be further decomposed into some subtasks or be an element task that is not decomposable; the way of decomposing a task can be fixed or multiple; the executing order of the tasks can be optional or should follow some precedence relationship defined between the tasks; the tasks can be fixed or be composed by tasks that arrive and depart over time.

(3) *Relations between the tasks and the agents*: Each agent can execute only one task at a time or some agents can execute multiple tasks simultaneously; each task requires exactly one agent to accomplish it or some tasks may require multiple agents; allocating tasks to agents with no planning for future allocations or each agent can be allocated several tasks which must be executed according to a given schedule.

(4) *Control models*: The system can use a centralized model that adopts a central controller that should know the status information of the entire system in real time and makes all decisions based on the information that is sent from the agents; or use a distributed control model where the tasks can be allocated by agents themselves and the global control unit is not needed, i.e., the agents make their decisions based on their own perceived information about the system; or use a hybrid control model

---

[1] An agent is rational means he is aware of his alternatives, forms expectations about any unknowns, has clear preferences, and chooses his action deliberately after some process of optimization (Osborne and Rubinstein, 1994b). In other words, a rational agents always acts to maximizes his utilities.

which can combine the centralized and decentralized models within a system.

(5) *Optimization objective*: The objective can be minimizing the response time of tasks, minimizing the makespan of tasks, maximizing the throughput of tasks, maximizing the reliability of tasks, maximizing the utility of task execution, and so on.

The above factors actually càn work as five axes which establish the problem space of task allocation nowadays widely studied. With respect to the basic problems in this space, techniques from combinatorial optimization and operations research suffice to give satisfactory solutions, but the fact is most problems in the space are NP-hard, so approximation algorithms studied in theoretical computer science and heuristic algorithms studied in artificial intelligence are useful; with respect to dynamic environments, where agents or tasks arrive and depart over time, it relates to online algorithms study; when time-extended assignment is allowed, it relates to planning; and when the agents are rational and privately hold some information we need, it relates to game theory and mechanism design. Actually, this is obviously an incomplete list; the basic intuition is that nowadays task allocation research has branched into a very large family that fundamentally correlates to combinatorial optimization and operations research, theoretical computer science, artificial intelligence, game theory and mechanism design, and so on.

In this chapter, we are not going to make a comprehensive survey on task allocation; actually, we think it is hard and unnecessary to do so, since task allocation has a very large and growing body of research work with various motivations, instead we focus on briefly summarizing several branches of task allocation research which is most interesting in our mind. Typically, these branches have considerably influenced the research on task allocation or have good value for further research. In particular, we will survey the following six important branches of task allocation research:

(1) *Taxonomy study on task allocation*, i.e., to study how to summarize and correlate the current study of task allocation and make a common taxonomy;

(2) *Allocating constrained or complex task*, i.e., to study how to allocate a set of tasks with complex interdependencies or multiple decomposition ways;

(3) *Task allocation for rational agents*, i.e., to study how to optimally allocate a set of tasks when the agents are rational players aiming to maximize individual utilities;

(4) *Task allocation for network systems*, i.e., to deal with the case where the agents are connected by an underlying network, which constrains the interactions of the agents;

(5) *Distributed task allocation*, i.e., to study how to allocate a set of tasks distributively without the need of a central controller;

(6) *Dynamic task allocation*, i.e., to deal with the settings with dynamic tasks or agents, for example, arriving or departing over time, changing over time, and so on.

We can discover from the literature that the above six directions are not only self-contained but also interrelated with each other. They collaboratively depict both the state-of-the-art and the development trends of task allocation research.

The remainder of this chapter is organized as follows: In Section 6.2, we summarize the work on taxonomy study on task allocation; in Section 6.3, we introduce the research on allocating constrained or complex tasks; in Section 6.4, we introduce task allocation for rational agents; in Section 6.5, we summarize the research on networked systems; in Section 6.6, we summarize the work on distributed task allocation; in Section 6.7, we introduce the work on dynamic task allocation; finally, we conclude this chapter with future work in Section 6.8.

6.2 Taxonomy Study on Task Allocation

Since task allocation is a rather general problem class that contains all the problems of the form that ask how to optimally allocate a set of tasks to a set of agents, a large amount of related research in the past decades can be put into this class. However, these research may differ from each other greatly on the details of the problem setting such as the properties of the agents, the properties of the task, the allocating method, and so on. Most of such research are aimed to solve the task allocation problem in a specific setting or focus on *ad hoc* or empirical work, with many coordination architectures having been proposed and validated in a proof-of-concept fashion, but infrequently analyzed (Gerkey and Matarić, 2004). It is of ut most importance to categorize the settings that have been investigated by current task allocation study, since it will not only give us a systematic and global overview on the problem domain and correlate existing studies but also motivate new research on the blind spots.

6.2.1 *Taxonomy Study for MultiRobot Task Allocation*

The most systematical taxonomy study for task allocation was done in 2004 by Gerkey and Matarić (2004) on the subfield of multirobot task allocation (MRTA), which categorizes MRTA problems based on optimization theory along the following three axes:

(1) *Single-task robots (ST) versus multitask robots (MT)*, distinguishes between problems in which each robot can execute only one task at a time and problems in which some robots can execute multiple tasks simultaneously.

(2) *Single-robot tasks (SR) versus multirobot tasks (MR)*, distinguishes between problems in which each task requires exactly one robot to achieve it and problems in which some tasks may require multiple robots.

(3) *Instantaneous assignment (IA) versus time-extended assignment (TA)*, distinguishes between problems concerned with instantaneous allocation of tasks to robots with no planning for future allocations and problems concerned with both current and future allocations, meaning that each robot is allocated several tasks which must be executed according to a given schedule.

Based on the above three axes and the two assumptions that all the tasks are independent from each other and the optimization objective can be defined as a utility function that specifies a utility for each feasible allocation, which intuitively capture the quality value and cost of each allocation, the problem space of MRTA can be divided into eight parts depicted as Table 6.1.

The above taxonomy is widely adopted in the literature and provides a common vocabulary for describing multirobot task allocation problems. However, it is restricted to independent tasks, and as such excludes a large collection of problems in the widely growing body of multirobot coordination work in which there are dependencies or constraints between tasks. In fact, interdependencies between the

Table 6.1. Gerkey and Matarić's taxonomy for MRTA

	SR	MR	
ST	ST–SR–IA is equivalent to the optimal assignment problem (OAP)(Gale, 1960) and is the only problem in this space that can be solved in polynomial time.	ST–MR–IA is equivalent to the set-partitioning problem. It is also referred to as coalition formation (Shehory and Kraus, 1995a), which is expressed as the problem of partitioning the set of robots into non-overlapping subteams to perform the given tasks.	IA
MT	MT–SR–IA is equivalent to the ST–MR–IA problem, with the roles of tasks and robots reversed.	MT–MR–IA is equivalent to the set-covering problem. It is a problem with a goal of trying to compute a coalition of robots to perform each task, where a given robot may be assigned to more than one coalition.	
ST	ST–SR–TA is a problem which involves determining a schedule of tasks for each robot. It is equivalent to the machine scheduling problem (Brucker, 2004).	ST–MR–TA is a problem that involves both coalition formation and scheduling.	TA
MT	MT–SR–TA is a problem mathematically equivalent to ST–MR–TA, but is less commonly studied.	MT–MR–TA is an extremely difficult problem that can be thought of as an instance of a scheduling problem with multiprocessor tasks and multipurpose machines.	

tasks is very common, for example, the basic multirobot routing problem (Lagoudakis *et al.*, 2005); moreover, there are a lot of tasks that can be decomposed into some subtasks in multiple ways (Zlot, 2006). Motivated by this, Korsah *et al.* (2013) developed an approach to differentiate these relationships between tasks based on the degree of interdependencies of agent-task utilities, which is not only a key distinguishing factor between different types of MRTA problems but also a strong determining factor of problem difficulty. Basically, problem features such as whether or not agents can execute more than one task at a time (ST versus MT agents) and whether tasks require one agent or multiple agents (SR versus MR tasks) can be translated into this. They proposed an improved taxonomy in which on the top the level provides descriptive information about the problem configuration utilizing Gerkey and Matarić's taxonomy, and a new level is introduced to define the degree of interdependence of agent-task utilities, which differentiates the following four classes:

(1) *No dependencies (ND)*: The effective utility of an agent for a task does not depend on any other tasks or agents in the system.

(2) *In-schedule dependencies (ID)*: The effective utility of an agent for a task depends on what other tasks that agent is performing. Constraints may exist between tasks on a single agent's schedule, or might affect the overall schedule of the agent.

(3) *Cross-schedule dependencies (XD)*: The effective utility of an agent for a task depends not only on its own schedule but also on the schedules of other agents in the system. For this class, allowable dependencies are "simple" dependencies in that the task decomposition can be optimally pre-determined prior to task allocation. Constraints may exist between the schedules of different agents.

(4) *Complex dependencies (CD)*: The effective utility of an agent for a task depends on the schedules of other agents in the system in a manner that is determined by the particular task decomposition that is ultimately chosen. Thus, the optimal task decomposition cannot be decided prior to task allocation, but must be determined concurrently with task allocation. Furthermore, constraints may exist between the schedules of different agents.

Table 6.2. Korsah *et al.*'s taxonomy for MRTA

Level 1	ND	ID	XD	CD
Level 2	ND[ST–SR–IA]	ID[ST–SR–TA]	XD[ST–SR–IA]	CD[ST–SR–IA]
	ND[ST–SR–TA]	ID[ST–SR–IA]	XD[ST–SR–TA]	CD[ST–SR–TA]
		ID[MT–SR–TA]	XD[MT–SR–IA]	CD[MT–SR–IA]
			XD[MT–SR–TA]	CD[MT–SR–TA]
			XD[ST–MR–IA]	CD[ST–MR–IA]
			XD[ST–MR–TA]	CD[ST–MR–TA]
			XD[MT–MR–IA]	CD[MT–MR–IA]
			XD[MT–MR–TA]	CD[MT–MR–TA]

Then in the second level, under each of the above four cases, there are a list of problems which can be further described by Gerkey and Matarić's taxonomy, and the proposed taxonomy is depicted by Table 6.2.

Note that some of the potential problems in the classes ND and ID, e.g., ND[MT–SR–IA], ID[MT–SR–IA], and so on, are missing. Korsah *et al.* (2013) argued that although the original Gerkey and Mataric taxonomy was meant for independent tasks and utilities, several of the original problems in their taxonomy do in fact have interrelated utilities. So, in the ND class, the only subclasses with meaningful problems are the ST–SR–IA and ST–SR–TA subclasses. All problems with MT and/or MR have some form of interrelated utilities and so are included in one or more of the ID, XD, and CD classes.

Compared with Gerkey and Matarić (2004), Korsah *et al.* (2013) obviously provide a more comprehensive taxonomy for MRTA. As has mentioned by Korsah *et al.* (2013), their taxonomy not only applies to multirobot systems, but also applies to humans or non-robotic vehicles, which they considered collectively to be agents. While both Gerkey and Matarić's work and Korsah *et al.*'s (2013) work aims to provide a global view to the field of MRTA, there is some work, e.g., Nunes *et al.* (2016) that aims to provide a more specific categorization of problems in a more narrow scope that has both temporal and ordering constraints. They proposed a taxonomy that builds on the existing taxonomy for multirobot task allocation and organizes the literature according to the temporal nature of the tasks.

6.2.2 *Taxonomy Study of Other Subfields*

Besides in the subfield of multirobot task allocation, taxonomy study is also conducted in other fields, but with relatively less originality or detail. For example, Lau and Zhang (2003) proposed a preliminary classification for task allocation via multiagent coalition formation based on three driving factors (demands, resources, and profit objectives). They divide their analysis into five cases. For each case, they presented algorithms and complexity results; Pizzocaro and Preece (2009) discussed the possibility to model multiple instances of the MultiSensor Task Allocation (MSTA) problem as specializations of the MRTA problem. They concluded that capturing essential characteristics of the MSTA problem requires an extension of the MRTA taxonomy to cover domain-specific features of sensor networks; this leads to the need for a new taxonomy of MSTA problems.

To sum up, a taxonomy study of task allocation aims to reduce the previously studied task allocation problems to basic mathematical problems, and correlate them in the bottom layer, and thus develops a global view of the studied problem space. A good taxonomy should typically be comprehensive and in-depth, that is, it should comprehensively include the studied problems and provide enough discrimination to distinguish them with each other. Although some taxonomy have been proposed for some subfields several years ago, they are becoming far and far from comprehensive and in-depth as time goes by, since task allocation is still a relatively active field and a lot of new settings or problems are discovered and studied over time. Moreover, there is still no taxonomy developed for general task allocation problems. Therefore, developing more updated and more general taxonomy for task allocation is of importance in the future research.

6.3 Allocating Constrained or Complex Tasks

The problems of allocating a set of basic tasks, in which each task is not decomposable and the execution of each task are independent from other tasks, are relatively simple and well-studied, especially in the early literature of task allocation. However, two variants of basic task are common in applications. Firstly, there is usually some constraints on or between the tasks. Secondly, some tasks may have to be further decomposed into subtasks before allocating, and the possible ways of decomposing can be multiple. These changes in the task set may lead to harder task allocation problems that need further study.

6.3.1 *Allocating Tasks with Constraints*

Allocation of tasks with constraints on when, where, and in what order they need to be done by groups of agents is an important class of problems with many real-life applications, including warehouse automation, pickup and delivery, surveillance at regular intervals, space exploration, search and rescue, and much more (Nunes *et al.*, 2016). Constraints may be defined on a task or between the tasks. The nature of constraints is that they can be potentially arbitrary functions restricting the space of feasible way of allocating the tasks. There are different types of constraints, and some most common ones are listed as follows (Korsah *et al.*, 2013):

- *Deadline constraints*: Some tasks must be finished within a deadline. For example, in search and rescue domains the tasks are discovered over time and have to be done as quickly as possible or in a specified deadline.
- *Simultaneity constraints*: Some tasks must be performed at the same time or may need to be done synchronously, as in surveillance where robots have to track multiple people at the same time.
- *Non-overlapping*: Some tasks cannot be performed at the same time, for example, you cannot make phone calls using your mobile phone when you are refueling your car in a gas station.
- *Precedence constraints*: Some tasks may need to be executed in a specific order, such as in urban disaster scenarios in which police must clear blockades from roads before firetrucks can find and put out fires.

- *Proximity constraints*: In problems where tasks have a choice of locations at which they can be performed, proximity constraints may specify that two tasks must be performed less (or greater) than a specified distance from each other.

Among these constraints, temporal constraints (e.g., the first four items) are most well-studied in the task allocation literature. Temporal constraints in a task set usually can be defined by specifying for each task a time window (Heilporn *et al.*, 2010; Koes *et al.*, 2005; Ponda *et al.*, 2012), which consists of a start time and a finish time. Moreover, a set of relationships can be specified between any two time intervals, e.g., after, before, equal, overlap, during, etc (Allen, 1990). An improved approach which is widely adopted in recent task allocation literature (e.g., Barbulescu *et al.*, 2010; Gombolay *et al.*, 2013; Hunsberger, 2002; Jr and Durfee, 2012; Nunes and Gini, 2015) is simple temporal network (STN) (Dechter *et al.*, 1991), which represents temporal constraints with a graph, in which nodes represent time point variables or time events, and weighted edges represent inequality constraints between time points. In a STN, constraint consistency can be efficiently verified in a polynomial time (Dechter *et al.*, 1991), and new time points and constraints can be dynamically added in a polynomial time, which is beneficial in dynamic domains where new tasks can appear and disappear (Nunes *et al.*, 2016).

The introduction of temporal and ordering constraints increases the c omplexity of task allocation, because solutions might contain assignments of tasks that depend on each other to different agents, creating execution dependencies among agents (Nunes *et al.*, 2016). Very naturally, for such tasks the best choice is time-extended allocation (TA), in which agents can maintain task sequences (Berhault *et al.*, 2003; Dias, 2004; Pongpunwattana *et al.*, 2006; Scerri *et al.*, 2005; Vidal, 2002; Zlot *et al.*, 2002) or schedules (Goldberg *et al.*, 2003; Lemaire *et al.*, 2004; Schneider *et al.*, 2005), or in some cases handle multiple tasks or roles concurrently (Farinelli *et al.*, 2003; Stroupe and Balch, 2011). A common technique for time-extended task allocation is for a system to have a central allocator that assigns all tasks to the team (Zlot, 2006). For example, Hussain *et al.*

(2003) describe a centralized approach to a capacitated multirobot routing problem in which a genetic algorithm determines routes and schedules for the team. However, in many cases, individual state information and agent capabilities are not available centrally and instead are distributed among the team. To avoid expensive (and potentially continuous) communication of this information, it is often preferable to compute cost estimates and scheduling decisions locally. Auction mechanisms are one manifestation of this technique (Dias *et al.*, 2006; Zlot, 2006). Since the robots have to reason about the cost dependencies between tasks, TA problems are often very demanding from a planning perspective. To relieve the high computational overhead, sometimes instantaneous allocation (IA) is also adopted for tasks with constraints. In IA approaches, each agent behaves myopically and only considers handling one task at any given time, ignoring many dependencies between tasks and potential upcoming commitments (Zlot, 2006). Dispatching is a commonly used TA approach in time-extended allocation domains (Fua and Ge, 2005; Gerkey and Matari, 2002; Godwin *et al.*, 2006; Kalra and Martinoli, 2006; Lin *et al.*, 2004a; Sariel *et al.*, 2008). By patching, one task is assigned to each available robot, and if there are more tasks than robots, the remaining tasks can be allocated once previous assignments have been completed. This approach is often adopted either for ease of implementation or in order to avoid the need for computationally expensive task sequencing or scheduling algorithms. Moreover, by keeping the planning horizon short, there is no need for rescheduling and a reduced need for task reallocation in situations of highly dynamic or uncertain environments. However, by ignoring many of the dependencies between task costs or utilities, poor solution quality can potentially occur since tasks are always assigned to the next available robots and not necessarily the globally best-suited ones (Zlot, 2006).

6.3.2 *Allocating Complex Tasks*

In some cases, a task can be decomposed into subtasks and be allocated to one or more agents. From this aspect, Zlot

(2006) proposed that tasks can be classified into the following categories:

- *Decomposition and decomposability*: A task t is decomposable if it can be represented as a set of subtasks t for which satisfying some specified combination (ρ_t) of subtasks in σ_t satisfies t. The combination of subtasks that satisfy t can be represented by a set of relationships ρ, that may include constraints between subtasks or rules about which or how many subtasks are required. The pair (σ_t, ρ_t) is also called a decomposition of t. The term decomposition can also be used to refer to the process of decomposing a task.
- *Multiple decomposability*: A task t is multiply decomposable if there is more than one possible decomposition of t.
- *Elemental task*: An elemental (or atomic) task is a task that is not decomposable.
- *Decomposable simple task*: A decomposable simple task is a task that can be decomposed into elemental or decomposable simple subtasks, provided that there exists no decomposition of the task that is multi(agent)-allocatable.
- *Simple task*: A simple task is either an elemental task or a decomposable simple task.
- *Compound task*: A compound task t is a task that can be decomposed into a set of simple or compound subtasks with the requirement that there is exactly one fixed full decomposition for t (i.e., a compound task may not have any multiply decomposable tasks at any decomposition step).
- *Complex task*: A complex task is a multiple decomposable task for which there exists at least one decomposition that is a set of multi[agent]-allocatable subtasks. Each subtask in a complex tasks decomposition may be simple, compound, or complex.

Korsah *et al.* (2013) summarize the features of a complex task as follows: A key difference between compound and complex tasks is that the optimal decomposition for compound tasks can be determined prior to task allocation, whereas for complex tasks, it is not known prior to task allocation which of the possible decompositions is optimal. Thus, a complete algorithm for allocating compound tasks can optimally decompose these into simple tasks prior to task allocation, whereas a complete algorithm for allocating complex tasks would need to explore the various possible task decompositions concurrently with task allocation. In addition to answering the basic task allocation question of "who does what?", an algorithm for allocating complex tasks also needs to answer the question "which simple tasks should be executed (or which decomposition should be used)?". The space of possible allocations for a multiagent task allocation problem with simple or compound tasks is exponential in the number of agents and tasks. The space of possible allocations for the same problem with complex tasks is exponentially larger than this.

In his seminal paper Zlot (2016) proposed a solution to complex tasks allocation which extends market-based approaches by generalizing task descriptions into task trees, thereby allowing tasks to be traded in a market setting dynamically at multiple levels of abstraction. In order to incorporate these task structures into a market mechanism, novel and efficient bidding and auction clearing algorithms are proposed. Explicitly reasoning about complex tasks presents a trade-off between solution efficiency and computation time. He analyzed the tradeoff for task tree auctions and further introduced a method for dramatically reducing the bidding time without significantly affecting solution quality.

There is also some follow-up work. Zheng and Koenig (2009) proposed a method for solving complex task allocation, where they develop a distributed negotiation procedure that allows robots to find all task exchanges that reduce the team cost of a given task allocation, without robots having to know how other robots compute their robot costs. Elmogy *et al.* (2009) proposed centralized and hierarchical dynamic and fixed tree task allocation approaches to solve complex task allocation problems in multisensor surveillance systems. The simulation results show that hierarchical dynamic tree task allocation outperforms all the other techniques, especially in complex surveillance operations where large number of robots is used to scan large number of areas. Jeyabalan *et al.* (2009) proposed a market-based approach for complex task allocation for wireless network based multirobot system. Landn *et al.* (2010) proposed a sound and complete distributed heuristic search algorithm for allocating the individual tasks in Task Specification Trees (TSTs) to platforms. The allocation also instantiates the parameters of the tasks such that all the constraints of the TST are satisfied. Constraints are used to model dependencies between tasks, resource usage, as well as temporal and spatial requirements on complex tasks. Zheng and Koenig (2011) proposed an auction-like algorithm for distributed task-allocation problems where cooperative agents need to perform some complex tasks simultaneously, in which each agent can construct and approximate a reaction function, which characterizes its costs of performing multiple complex tasks, in a distributed way. Khamis *et al.* (2011) studied how to optimally assign a set of surveillance tasks to a set of mobile sensing agents to maximize overall expected performance, taking into account the priorities of the tasks and the skill ratings of the mobile sensors. This chapter presents a market-based approach to complex task allocation, in which both centralized and hierarchical allocations are investigated as winner determination strategies for different levels of allocation and for static and dynamic search tree structures.

6.4 Task Allocation for Rational Agents

For some MAS such as crowdsourcing systems, social networks, and so on, the agents are typically human which are usually rational in the sense of game theory and hold some information we need as private information. For example, the quality value, cost value of executing a task, and the probability of being able to accessing a resource are parameters privately held by the agents. However, these parameters are crucial in designing the optimization algorithm. In other words, with the uncertainties in these parameters we are not sure about which allocation is the best.

A possible approach for solving this issue is to ask the agents to report(or bid) these required private parameters, which can also be called their *types*, before performing task, allocation; then after obtaining the reports of the agents, we can select an optimal allocation and pay each agent an amount that is sufficient to compensate her cost. But the problem is each agent may not necessarily truthfully report their parameters, they will only report in a way that maximizes their utility. Therefore, the rules for allocating and paying should be carefully designed to at least satisfy the following conditions:

(1) *Incentive compatible*: Truthful reporting is the best choice of all the agents. In particular, when the probability distribution of each agent's type is available, every agents truthfully reporting their types can be a Bayesian–Nash equilibrium; otherwise, in the case of prior-free, truthfully reporting can be each agent's dominant strategy. Incentive compatible is sometimes also called *truthful*.

(2) *Individually rational*: Every agent will be guaranteed a non-negative utility. When incentive compatibility is satisfied, we only have to make sure the truthfully reporting utility of each agent is non-negative.

(3) *Computationally tractable*: The allocation and payment rules should be able to be computed in polynomial time.

(4) *Constant-factor approximative*: The selected allocation should be constant-factor approximate the optimal allocation.

The above methodology actually solves the task allocation problem in the strategic case in the framework of *algorithmic mechanism design* (Nisan and Ronen, 1999, 2001), which is nowadays extensively studied in some communities of theoretical computer science and MAS. In particular, task allocation mainly relates to the following important directions of algorithmic mechanism design:

- *VCG-based mechanism design* (Babaioff and Blumrosen, 2008; Buchfuhrer *et al.*, 2010; Hartline, 2010; Nisan *et al.*, 2004). To design computationally feasible mechanisms with good performance guarantee and good game theoretical incentives based on Vickrey-Clarke-Groves (VCG) mechanisms.
- *Optimal mechanism design* (Cai *et al.*, 2013a, 2013b; Hartline and Karlin, 2007a; Li and Yao, 2013; Myerson, 1981a; Singer, 2010). To optimize objective functions in mechanism design,

especially that relates to payments, without suffering from the good game theoretical properties such as incentive compatibleness and incentive rationality.

- *Online mechanism design* (Friedman and Parkes, 2003; Hartline and Karlin, 2007b; Lavi and Nisan, 2004, 2005; Mashayekhy *et al.*, 2016). To design mechanisms for bidders arriving over time, with the performance competitive to the offline case, just like that in online algorithms.

In fact, the above three directions intrinsically interrelate with each other. There are a lot of research which solve task allocation in the strategical case based on methodologies from one or more directions above.

6.4.1 *Task Allocation via VCG-based Mechanisms*

VCG mechanism (Clarke, 1971; Groves, 1973; Vickrey, 1961) is one of the most famous positive results in mechanism design. It is a truthful mechanism for maximizing social welfare, which can be defined as the sum of the values of all the agents. Therefore, VCG mechanism is suitable for task allocation problems with the optimization objective of maximizing the overall value of all the agents. However, since finding the optimal allocation (even in the complete information case) is usually intractable, tractable constant-factor approximation mechanisms that maintain incentive compatibility (IC) and individually rationality (IR) are useful substitutes. But finding an approximate mechanism is typically more tricky than finding an approximation algorithm, which is well studied in theoretical computer science, since we have to, at the same time, properly design the payment function to maintain IC and IR.

In the seminal paper of algorithmic mechanism design, Nisan and Ronen (2001) carefully studied the following task allocation problem: Given k tasks that need to be allocated to n agents, each agent i's type is, for each task j, the minimum amount of time t_{ij} in which it is capable of performing this task. The goal is to minimize the completion time of the last assignment (the make-span). They studied this problem under two main models: a basic mechanism design-based model and a model that allows more information to be incorporated into the mechanism, or be called *mechanisms with verification*, which takes advantage of the fact that computers actually act (execute a task, route a message, etc.) to gain extra information about the agents types and actions, i.e., for the task allocation problem we can assume that by the end of the execution the mechanism knows the exact execution time of each task. Under the assumptions of the basic model, they showed that the problem cannot be approximated within a factor of $2 - \epsilon$. Then, under the second model assumptions, they introduced several novel mechanisms including optimal, constrained optimal, and polynomial-time approximation mechanisms. They have also shown that worst-case behavior can be improved using randomness without weakening the "game-theoretic" requirements of the mechanism. Mechanisms with verification were further studied by Zhao *et al.* (2016b) on a general task allocation problem where multiple agents collaboratively accomplish a set of tasks, but they may fail to successfully complete tasks assigned to them. To design an efficient task allocation mechanism for this problem, they showed that post-execution verification based mechanism is truthfully implementable if and only if all agents are risk-neutral with respect to their execution uncertainty. They also showed that trust information between agents can be integrated into the mechanism without violating its properties, if and only if the trust information is aggregated by a multilinear function. It is the first time that this characterization and bound of the applicability of the post-execution verification based mechanism has been studied. They further demonstrated the applicability of the mechanism in the real world and showed the significance of the post-execution verification in the design. Weerdt *et al.* (2012); Zhang and Weerdt (2007) studied the problem of finding truthful mechanisms for social task allocation problems. In this chapter they give on the one hand an optimal mechanism and model the problem as an integer linear program (ILP), and on the other hand give a polynomial-time approximation by splitting the problem into smaller subproblems, each of which is solved optimally.

6.4.2 *Task Allocation Based on Optimal Mechanisms*

Optimal Mechanism design was initially studied by Myerson (1981a) in the Bayesian case, where the distribution of each agent's type is public information. Myerson (1981a) has proposed a Bayesian optimal mechanism for the case of selling only one item to a set of agents, i.e., each agent has only one unknown parameter, while for the general case, where each agent has more than one unknown parameter, it is left open. Although this problem has been in-depth studied by a lot of researchers during the past decades, there is little substantive progress. The best results were obtained by Cai *et al.* (2013a, 2013b) a few years ago, which have proposed a blackbox reduction from Bayesian optimal mechanism design to algorithm design for general optimization objective functions. But their approach is limited to the additive case; in the general case, the problem is still widely open. For the prior-free case where the type distribution of each agent is unavailable, the optimal mechanism design problem is even harder, but there is some positive results on some subcases, for example, budget-feasible mechanisms (Anari *et al.*, 2014; Bei *et al.*, 2012; Chen *et al.*, 2011; Singer, 2010) and competitive mechanisms for digital goods (Chen *et al.*, 2014; Goldberg *et al.*, 2006, 2001; Kennes, 2006).

Among the class of optimal mechanisms, budget-feasible mechanisms have been used in many task allocation problems, especially crowdsourcing task allocation (Anari *et al.*, 2014; Biswas *et al.*, 2015; Zhao *et al.*, 2014, 2016a). Intuitively, budget-feasible mechanisms are the class of mechanisms which maximize the value of the allocation with the total payment under a predefined budget for reverse auctions, in which the auctioneer purchases items from the bidder with a demand valuation function specifying the value of each bundle of items. Current budget-feasible mechanisms are built on the single-parameter model (Myerson, 1981a): in single parameter domains a normalized mechanism is truthful iff the allocation function is monotone; and winners are paid threshold payment. In the seminal paper (Singer, 2010), Singer shows that when the demand valuation function is a monotone submodular function, we can find out a computationally feasible mechanism with constant-factor approximation guarantee. This result is adopted and improved in some research on crowdsourcing task allocation. Anari *et al.* (2014) consider a mechanism design problem in the context of large-scale crowdsourcing markets such as Amazon's Mechanical Turk, ClickWorker, CrowdFlower. In these markets, there is a requester who wants to hire workers to accomplish some tasks. Each worker is assumed to give some utilities to the requester. Moreover, each worker has a minimum cost that he wants to get paid for getting hired. This minimum cost is assumed to be private information of the workers. The question then is: if the requester has a limited budget, how to design a direct revelation mechanism that picks the right set of workers to hire in order to maximize the requester's utility. They showed that this setting can be solved by the framework of budget-feasible mechanism design, and in such a large market, the approximation guarantee can be further improved. Biswas *et al.* (2015) studied a multiarmed bandit (MAB) problem with several real-world features: the requester wishes to crowdsource a number of tasks but has a fixed budget which leads to a trade-off between cost and quality while allocating tasks to workers; each task has a fixed deadline and a worker who is allocated a task is not available until this deadline; the qualities (probability of completing a task successfully within deadline) of crowd workers are not known; and the crowd workers are strategic about their costs. They proposed a mechanism that maximizes the expected number of successfully completed tasks, assuring budget feasibility, incentive compatibility, and individual rationality. Zhao *et al.* (2014, 2016a) design two mechanisms, OMZ and OMG, satisfying the computational efficiency, individual rationality, budget feasibility, truthfulness, consumer sovereignty, and constant competitiveness under the zero arrival–departure interval case and a more general case, respectively.

6.4.3 *Task Allocation Based on Online Mechanisms*

Online mechanisms extend the methods of mechanism design to dynamic environments with multiple agents and private information. Decisions must be made as information about types is revealed online and without knowledge of the future, in the sense of online algorithms (Hartline and Karlin, 2007b). Since many task allocation applications face dynamic environments, there are a lot of task allocation studies based on online mechanisms.

Wang *et al.* (2016) proposes an incentive mechanism with privacy protection in mobile crowdsourcing systems. Combining the advantages of offline incentive mechanisms and online incentive mechanisms, they proposed an incentive mechanism that selects the worker candidates statically, and then dynamically selects winners after bidding. Zhao *et al.* (2014, 2016a) designed online incentive mechanisms to motivate mobile users to participate in mobile crowd sensing; they have proven that their mechanisms satisfy the computational efficiency, individual rationality, budget feasibility, truthfulness, consumer sovereignty, and constant competitiveness. Besides, they showed their main idea and framework can also be applied to designing online incentive mechanisms with other objectives such as frugality and profit maximization. Chandra *et al.* (2015) considered the following setting: a requester has a set of tasks that must be completed before a deadline; agents (aka crowd workers) arrive over time and it is required to make sequential decisions regarding task allocation and pricing. Agents may have different costs for providing service, and these costs are private information of the agents. They assume that agents are not strategic about their arrival times but could be strategic about their costs of service. In addition, agents could be unreliable in the sense of not being able to complete the assigned tasks within the allocated time; these tasks must then be reallocated to other agents to ensure on-time completion of the set of tasks by the deadline. For this setting, they propose two mechanisms, both of which are dominant strategy incentive compatible, budget feasible, and also satisfy *ex post* individual rationality for agents who complete the allocated tasks. Jain *et al.* (2014) proposed a MAB incentive mechanism for crowdsourcing demand response in smart grids. Zhu *et al.* (2014) and Feng and Zilberstein (2004) proposed two truthful auction mechanisms for two different cases of mobile crowdsourcing with dynamic smartphones. They showed the proposed auction mechanisms achieve truthfulness, individual rationality, computational efficiency, and low overpayment.

6.5 Task Allocation for Networked Systems

There is also a large body of work on task allocation for networked systems, in which there are underlying networks between the agents, for example, social networks, wireless sensor networks, and so on. The nature of these networks is constraints between the agents on their cooperation, communication, and so on.

6.5.1 *Task Allocation in Social Networks*

Social networks are common entities in the world. The studies on social networks can be dated back to the 1930s. In its more than 80 years' development process, it continuously penetrates into the research domains of physics, computer science, economics, and other fields and has made a lot of important achievements. In particular, the emergence of online social networks, such as Facebook, Twitter, Weibo, Weixin, etc., unprecedentedly connect massive users together to become an effective platform for information dissemination, product marketing, and advertising and to play an increasingly important role. A social network can typically be modeled as a graph in which the nodes represent the users(or agents), and the links between the nodes are defined by a social relation.

In many social networks, social individuals often need to work together to accomplish a complex task (e.g., software product development, complex system testing, big data collecting, etc.). Task allocation in a social network has to consider the constraints of social relations. Weerdt *et al.* (2007, 2012) studied a case where each agent has a limited amount of resources of different types at her disposal; but each agent can cooperate with her neighbors by sharing some of her resources; so when an agent is allocated a task which requires some resources for execution, she can obtain some required resources from her neighbors in order to execute this task. Task allocation in this setting can be solved based on the manager/contractor architecture (Jiang and Jiang, 2009; Jiang *et al.*, 2013a, 2013b; Weerdt *et al.*, 2007), where a manager agent is allocated to a task using a centralized heuristic and then negotiates with the other contractors for resource assistance using a distributed heuristic (Jiang *et al.*, 2013b). Actually, the manager/contractor architecture is inherited from contract net protocol (Smith, 1980), which is a classical task allocation method proposed in the early 1980s. The task allocation process based on manager/contractor architecture can be described as follows: A task may be first allocated to one agent, which takes charge of the execution of the task (this agent is called the manager agent). When the manager agent lacks the necessary resources to execute the allocated task, it negotiates with its neighbors in the social network; if some neighbors have the required resources (the agents that provide resources to the manager agent is called contractor agents), the manager and contractor agents will work together to execute the task. In some cases, cooperation in a wider range is also possible, for example, Wang and Jiang Wang and Jiang (2014) studied an alternative case where the agents' cooperation domain is constrained in community (rather than neighbors) and each agent can negotiate with its intra-community member agents.

Basically, task allocation is the problem of finding out the allocation with the maximum utility. Weerdt *et al.* (2007, 2012), defined utility as the sum of all the values of the tasks which can be successfully executed. But in some cases, the execution time of the tasks is also an important factor in evaluating the performance. To reduce the execution time of a task, one of the key problems is to reduce the time used accessing the resources necessary for the task, i.e., *resource access time*, which include two factors: the communication time between the manager agent and contractor agents in the social network, and the tasks waiting time for resources at the agents (Jiang and Huang, 2012). To reduce the system resource access time, Jiang and Jiang (2009) presented a contextual resource negotiation mechanism by allowing agents to negotiate with others from nearby to faraway gradually. To achieve dependable resources with the least resource access time for undependable social networks, Jiang *et al.* (2013b) propose a reputation-based negotiation mechanism.

A network layer oriented task allocation model is presented for minimizing the task execution time in multiplex networks (Jiang *et al.*, 2015). A similar optimization factor considered in the literature is social effectiveness (Anagnostopoulos *et al.*, 2012; Kargar and An, 2011; Lappas *et al.*, 2009; Majumder *et al.*, 2012; Rangapuram *et al.*, 2015; Wang and Jiang, 2015), which is a useful indicator to determine how effectively social individuals can communicate. Wang and Jiang (2015) studied a case where a set of interdependent tasks have to be allocated to a team which not only can execute this task set but also has satisfactory social effectiveness.

Since social networks typically consist of large amounts of nodes, centralized allocation approaches are often not feasible due to the fact that global information cannot be gathered in real time and the central controller may become a performance bottleneck of the system (Jiang, 2015). Distributed frameworks are often more suitable for this setting. For example, the work of Weerdt *et al.* (2012) assumes that agents have only local information about tasks and resources, and so proposes a distributed task allocation algorithm; Wang and Jiang (2015) presents a distributed multiagent-based task allocation model by dispatching a mobile and cooperative agent to each subtask of each complex task, which also addresses the objective of social effectiveness maximization. Moreover, the widely adopted manager/contractor architecture, where firstly tasks is allocated to managers by a centralized approach and then the managers negotiate with the contractors on cooperation distributively (Jiang *et al.*, 2013b), is intrinsically a hybrid framework.

Moreover, an interesting phenomenon in social networks is that properties of network topology have notable influence on the performance of task execution. Gaston and Desjardins (2005, 2008) showed that the agents interaction topology is a key factor in effective team formation, mainly that scale-free networks support the diversity required for effective teams. Gaston and Desjardins (2005, 2008) and Kota *et al.* (2009) thus developed a structural adaptation method to increase social welfare, where agents can adjust the network structure by deleting their costly connections and rewiring them to those agents that have better connections. Weerdt *et al.* (2012) also investigated how the different network topologies affect the performance of the system.

Finally, since social network nodes are typically human or belong to different interest group, they are usually rational in the sense of game theory. In some work (e.g., Chevaleyre *et al.*, 2007; Sandholm, 1998), connected agents are allowed to negotiate about exchanging resources locally in a network, which will converge to new allocations with properties such as envy-freeness (Amador *et al.*, 2014). Resource allocation in a social network has also been realized by letting each pair of connected agents solve a bargaining problem (Chakraborty *et al.*, 2009; Kleinberg, 2008), which can be thought of as sharing a single pie among the two agents. Results show, for example, existence and computation of equilibrium in specific graph structures, given efficiency (and fairness) criteria such as the Nash bargaining solution (Weerdt *et al.*, 2012). When individual rationality and information incompleteness are taken into consideration, task allocation in social networks can be solved based on the framework of mechanism design (An *et al.*, 2010; Weerdt *et al.*, 2012; Ye *et al.*, 2013), which intuitively solves the task allocation optimization problem in uncertainty environments based on truthfully eliciting the required unknown information from the agents.

6.5.2 *Task Allocation in Wireless Sensor Networks*

Since sensors in a wireless sensor network (WSN) typically have limited resources of energy, computation, storage, communication bandwidth, and so on, sensors are significantly resource-constrained devices and last till the depletion of their batteries (Younis *et al.*, 2003). It is important to design its task allocation algorithm reasonably to reduce resource consumption. For example, Yu and Prasanna (2005) proposed an energy-balanced allocation of a real-time application onto a single-hop cluster of homogeneous sensor nodes connected with multiple wireless channels, in which the time and energy costs of both computation and communication activities are considered. Edalat *et al.* (2009) focused on task allocation in WSNs, which is performed with the aim of achieving a fair energy balance among the sensor nodes while minimizing delay using a market-based architecture. Li *et al.* (2010) studied task allocation in cluster-based WSN using negotiation model with two goals: firstly, finding out energy efficiency task allocation; and secondly, maintaining energy balance of nodes in WSN after the task to prolong the network lifecycle. They proposed a new contract net-based task allocation model for WSN and created an energy consumption based utility function to evaluate the allocation plan. Simulation results show that the proposed improved mechanism of MAS auto-negotiation model is a real time and effective scheme, which can be carried out to allocate tasks for WSN with a high success rate. Chen *et al.* (2012) designed a task allocation for WSN based on contract net with the aim of reducing energy consumption and traffic flow. Je *et al.* (2012) proposed a heuristic repeated matching task allocation methodology for designing the secure in-vehicle network to reduce high computational complexity. Compared to the typical task allocation problems in the in-vehicle control network, the proposed task allocation problem considers the varying probability of task vulnerability exploits as a new objective function. Younis *et al.* (2003) pointed out that most of the previous research focused on the optimal use of sensor's energy, very little attention has been paid to the efficiency of energy usage at the gateway, and presented an optimization scheme for task allocation to gateways.

Other optimization objectives include coverage or performance. For example, Low *et al.* (2004, 2006) present a task allocation scheme by self-organizing swarm coalitions for distributed mobile sensor network coverage. They proposed an approach to self-regulate the regional distributions of sensors in proportion to that of the moving targets to be tracked in a non-stationary environment. As a result, the adverse effects of task interference between robots are minimized and sensor network coverage is improved. Quantitative comparisons with other tracking strategies show that their approach can provide better coverage and greater flexibility to respond to environmental changes; Mi *et al.* (2014) present a novel threshold-based task allocation strategy to effectively assign tasks. Yang *et al.* (2014) proposed a modified version of binary particle swarm optimization for allocating tasks to sensors, where multiple metrics, including task execution time, energy consumption, and network lifetime, are considered as a whole by designing a hybrid fitness function to achieve the best overall performance.

6.5.3 · *Other Researches for Networked Task Allocation*

Besides work on social networks and WSN, there are also some studies on task allocation for other networked MAS, for example, network of processors (Hsu *et al.*, 2001), grids (Liu *et al.*, 2005b), or directly the abstract model of agent network (Abdallah and Lesser, 2005; Eberling and Büning, 2010; Kafal and Yolum, 2012; Shehory, 1999; Vidal, 2003). With respect to these systems, the problem of how to deal with the restrictions on the cooperation of the agents resulting from the underlying network is still a basic research topic. For example, Eberling and Büning (2010) studied the problem of task allocation in a MAS with agents having only local knowledge, in which agents can cooperate with neighbors to perform assigned tasks. They proposed a local learning algorithm that favors the decision to cooperate and so the agents adapt to the best neighbor and reach high levels of cooperation. In a similar setting where agents are not fully aware of all other agents and their demand for resources, Kafal and Yolum (2012) investigated how different communication protocols improve the efficiency of the resource allocation. Networks can also be used to model limited interactions between agents and mediators, which in this context are agents who receive the task and have connections to other agents. Abdallah and Lesser (2005) studied the decision process of mediators on breaking the task up into subtasks and negotiating with other agents to obtain commitments to execute these subtasks.

6.6 Distributed Task Allocation

Task allocation can be achieved based on centralized models, where there are central controllers that should know the status information of the entire system in real time and make all decisions based on the information that is sent from other nodes, or based on distributed models, where there is no central control unit and the tasks can be allocated by nodes themselves. Centralized models can be implemented relatively easier; however, they may be sometimes infeasible in real distributed systems that are large and dynamic, where the global information cannot be achieved in real time; moreover, the central controller may become a performance bottleneck of the system. In contrast, distributed models can be used in those large and dynamic distributed systems where the nodes make their decisions based on their own perceived information about the system and fault tolerance can be achieved because each node can act as the allocator (Jiang, 2015). There are a lot of approaches to distributed task allocation. We focus on summarizing three of them, which we feel most interesting, that is, the contract net protocol, distributed task allocation based on market-based approaches, and distributed task allocation based on coalition formation.

6.6.1 *The Contract Net Protocol*

A well-known distributed task allocation mechanism proposed by Smith (1980) in the early 1980s is the Contract Net Protocol (CNP), which is a task-sharing protocol in MAS discussed in the context of distributed problem solving (DPS). Basically, by CNP an agent, called a manager, that attempts to execute a task may divide it into several subtasks and contract each subtask to another agent, called a contractor, by a decentralized negotiation processes. Moreover, a contractor who receives a subtask can execute it or turn into a manager and restart the decomposing and negotiating process and further subcontract the sub–subtasks to other contractors. Finally, the managers and contractors forms a *contract net* which collaboratively solves the given task distributively. Note that a manager is responsible for monitoring the execution of a task and processing the results of its execution while a contractor is responsible for the actual execution of the task. Manager and contractor are roles that any node can take dynamically during the course of problem solving, instead of one being designated *a priori* as a manager or a contractor. An agent will typically often take on both roles simultaneously for different contracts. So, agents are not statically tied to a control hierarchy and a sense of efficient utilization of the agents' power can be achieved. A contract is established by a process of local mutual selection based on a two-way transfer of information: firstly, available contractors evaluate task announcements made by several managers and submit bids on those for which they are suited; then the managers evaluate the bids received for each subtask, and send a offer to the the winner which they determine to be most appropriate; finally, the bidders decide on to accept or to reject the offers they have received. In some cases, a contractor may restart the negotiation process and further partition a task and award contracts to other agents and becomes the manager for those contracts. This leads to the hierarchical control structure that is typical of task-sharing. In the above model, control is distributed because processing and communication are not focused at particular nodes, but rather every node is capable of accepting and assigning tasks.

Since it was proposed in the 1980s, CNP is extensively studied and widely adopted as a powerful coordination mechanism for MAS. Nowadays, it is still a popular methodology and is applied in a lot of most recent work (Jiang and Jiang, 2009; Jiang *et al.*, 2013a, 2013b; Weerdt *et al.*, 2007). Moreover, CNP has been continuously evolving in the past decades (Xu and Weigand, 2001). Many variants of the CNP have been developed for different problem domains, and many improvements or extensions for CNP have been proposed, for example, Davis and Smith (1983) improved the original framework of CNP to handle temporal constraints and to deal with scheduling conflicts. Sandholm (1993) pointed out that the bidding and award-ing decision process was left undefined in the original contract net task allocation protocol, and so fixed this problem by formalizing this process based on marginal cost calculations based on local agent criteria. In addition, CNP is extended to allow for clustering of tasks, to deal with the possibility of a large number of announcements and bid messages and to effectively handle situations, in which new bidding and awarding is being done during the period when the results of previous bids are unknown. By focusing on the issue of self-interest agents in automated negotiation system, Sandholm and Lesser (1996) extended CNP to Leveled Commitment Contract, which allows agent to end an ongoing contract if an outside offer is more profitable than current contract (Xueguang and Haigang, 2004). FIPA (2001) provides a well-defined model of how agents interact in negotiations, and has been widely applied and further extended.

Actually, CNP can be seen as an (rudimentary) example of market-based approaches to task allocation (Dias *et al.*, 2006).

6.6.2 *Market-based Distributed Task Allocation*

In a market-based task allocation, tasks are allocated to the agents using technologies and metaphors from economics or by a market mechanism. Market-based approaches mainly have two functions in solving the task allocation problem: firstly, they are well-studied methodologies for reliably finding out the optimal allocation in the strategic case where the agents are rational in the sense of game theory and hold some information we need as private information. Against this background, market-based techniques are attractive because their point of departure is developing protocols that achieve good system wide properties despite the fact that the agents act selfishly (Clearwater, 1996) (we have summarized this aspect in Section 6.4); secondly, some market mechanisms provide very good decentralized control models for task allocation; we will briefly introduce this aspect in this subsection.[2]

In general, there are mainly three classes of market mechanisms that can be adopted in task allocation such as

(1) *Combinatorial auction*, which models resource-based task allocation, where the agents with resources are the auctioneer(s) and

---

[2]Notice that a lot of task allocation studies, especially in the robotics community, using market-based approaches are based on this motivation and agents are assumed to be designed truthful. So, they only need to adopt simple untruthful auction mechanisms without the need of bothering to design the payment function. This is a key difference to the auctions nowadays studied in economics or theoretical computer science.

the agents who want to obtain the resources (for completing their tasks) are bidders;

(2) *Reverse auction*,[3] which models the case where the auction-eer (who has task(s)) procure resources or services from the bidders;

(3) *Continuous double auction*, which models the case where buyers and sellers continuously submit bids (an offer to buy at price p_b) and asks (an offer to sell at price p_a), respectively, (which are listed on a billboard), and the market clears (i.e., a transaction occurs) whenever the bid of a buyer matches the ask of a seller (i.e., when $p_b \geq p_a$).

Note that combinatorial auction and reverse auction are intrinsi-cally centralized methods since they involve centralized mechanisms in which bidders report their values to a center (the auctioneer) which then decides on the optimal allocation and the payments. In contrast, continuous double auction is intrinsically a decentralized method in which the allocation of the tasks are not computed by any single agent, but rather emerges out of the interactions of the agents in the protocol (Dash *et al.*, 2007).

Although combinatorial auction and reverse auction are central-ized models, they can be adopted to develop partial-decentralized (or hybrid) task allocation model in combine with the manager/ contractor model from the contract net protocol, in the sense that a manager agent is allocated to a task using a centralized heuristic and then negotiates with the other contractors for resource assistance using a distributed heuristic (Jiang *et al.*, 2013a, 2013b). In some settings, for example in a social network where each of the tasks are initially located in different agents (managers), each manager can locally cooperate with some contractors in task execution based on an auction, and this model can be seen as a decentralized model (Weerdt *et al.*, 2007, 2012). Moreover, by a combinatorial auction

---

[3]In an ordinary auction, the auctioneer sells items to the bidders, while in a reverse auction, the auctioneer buy items or services from the bidders. So, reverse auction is also called procurement auction.

(or reverse auction) we can distribute much of the planning and execution over the team and thereby retain the benefits of distributed approaches, including robustness, flexibility, and speed (Dias *et al.*, 2006). For example, in a multirobot sensing problem, where we have some robots located in different locations on the Mars and we are asked to sensing some location efficiently. By auction, we can ask each robot to plan its path to the location distributively and bid its path length (cost), and then allocate this sensing task to the robot with the lowest cost.

Continuous double auction is a more complete decentralized model for task allocation. Nevertheless, despite this decentralization, continuous double auctions still produce solutions that are very close to the optimal, even when the participants adopt very simple strategies (Gode and Sunder, 1993). For example, Dash *et al.* (2007) proposed a fully decentralized task allocation mechanism based on a continuous double auction which extend the standard form of this protocol (Gode and Sunder, 1993) by introducing a novel clearing rule, and empirically demonstrate (with simple trading strategies) that the proposed mechanism achieves high efficiency. Izakian *et al.* (2010) introduced a continuous double auction method for grid resource allocation in which resources are considered as provider agents and users as consumer agents. Garg *et al.* (2013) proposed a novel continuous double auction-based meta-scheduler mechanism that schedules parallel applications on global grids providing improvements that benefit both users and resources in terms of their effective utilization.

6.6.3 *Distributed Task Allocation via Coalition Formation*

Basically, task allocation via coalition formation asks the problem: how do we form teams of agents (coalitions), with each team assigned to a particular task, in order to best complete the set of tasks at hand? (Service and Adams, 2011): Notice that a marked feature of this approach is it allocates tasks to coalitions instead of individual agents, which is fundamentally different from some other popular methods, e.g., the CNP, in which tasks are allocated to single agents and a procedure for task-partitioning is necessary. In contrast, coalition formation solves the problem of assigning tasks to groups of agents. Such a solution is necessary in cases where single agents cannot perform tasks by themselves and tasks cannot be partitioned, or the partition is computationally too complex (Shehory and Kraus, 1998).

Coalitions are initially research objects of coalitional game theory, which concentrates on checking the stability (or fairness) and calculation of the corresponding payments for a previously formed coalitional configuration (that is, a partition of the agents to subsets). However, coalitional game theory does not provide algorithms which agents can use in order to form coalitions. Therefore, designing effective supporting algorithms is the key point for developing a task allocation method based on this theory. The coalition formation problem is equivalent to some classic optimization problems: if each agent can take part in more than one coalitions at the same time, i.e., overlapping coalitions is permitted, coalition formation can be formalized as a set covering problem, otherwise it can be formalized as a set partition problem. It is well known that both of their two problems are NP-complete. Based on both the algorithmic aspects of combinatorics and approximation algorithms for NP-hard problems, in the seminal papers, Shehory and Kraus (1995b, 1998) present efficient distributed coalition formation algorithms with low ratio bounds and with low computational complexities.

Shehory and Kraus's work was extended and improved by a lot of follow-up research. For example, Sandholm *et al.* (1999) proposed that coalition formation should include three activities: coalition structure generation, solving the optimization problem of each coalition, and dividing the value of the generated solution among agents, and since Shehory and Kraus's work focuses on cooperative rather than self-interested agents, payoff distribution is a non-issue and is thus not addressed, i.e., they actually only solved the coalition structure generation problem. Moreover, the approximation ratio bound is improved in Sandholm *et al.* (1999)'s algorithms, that is, they can establish a tight bound within minimal amount of search, and show that any other algorithm would have to search strictly more. Service and Adams (2011) proposed some coalition formation-based task allocation algorithms which provide guarantees on utility rather than solution cost, which is considered in Shehory and Kraus's algorithms. Their algorithm for general multiagent domains is a modification of and has the same running time as Shehory and Kraus algorithm. Tosic and Agha (2004) proposed an algorithm for distributed coalition formation based on a distributed computation of (maximal) cliques of modest sizes in the underlying communication network of agents. To summarize the research work on coalition formation-based task allocation, Lau and Zhang (2003) proposed a preliminary classification for the coalition formation problem based on three driving factors (demands, resources, and profit objectives).

6.6.4 *Centralized and Distributed Model: Trade-offs*

Being decentralized is a double-edged sword. Fully centralized approaches employ a single agent to coordinate the entire team. In theory, this agent can produce optimal solutions by gathering all relevant information and planning for the entire team. In reality, fully centralized approaches are rarely tractable for large teams, can suffer from a single point of failure, have high communication demands, and are usually sluggish to respond to local changes. In fully distributed systems, agents rely solely on local knowledge. Such approaches are typically very fast, flexible to change, adapt for varying system scales, and are robust to failures, but can produce highly suboptimal solutions, since local solutions may not necessarily aggregate to a good global solution (Dias *et al.*, 2006). Moreover, the negotiation among nodes may produce additional computation costs (Jiang, 2015). So, centralized approaches are most suited for applications involving small teams and static environments or easily available global information; fully distributed approaches are most suited for applications where large teams carry out relatively simple tasks with no strict requirements for efficiency (Dias *et al.*, 2006). In fact, a vast majority of coordination approaches have elements that are centralized and decentralized and thus reside in the middle of the spectrum. Sometimes those approaches can be called hybrid control model (Jiang, 2015). Market-based approaches fall into the hybrid category, and in some instances, they can produce more centralized or more distributed solutions. Approaches that adopt a hierarchical control model, which is intrinsically inherited from the CNP (Abdallah and Lesser, 2005; Jiang and Jiang, 2009), also fall into this hybrid category.

6.7 Dynamic Task Allocation

Dynamic task allocation in MAS extends the problem of task allocation to a more complex setting where agents or tasks will arrive and leave dynamically and the future is uncertain. In such settings, decisions must be made without information of the agents or tasks that have not arrived yet, coupled perhaps with uncertainty about which decision will be feasible in the future periods. So, dynamic properties come from tasks or agents, and based on this, dynamic task allocation problems can be classified into three types: allocating dynamic tasks to static agents, allocating static tasks to dynamic agents, and allocating dynamic tasks to dynamic agents.

6.7.1 *Allocating Dynamic Tasks*

Basically, dynamic tasks may mainly refer to two settings: firstly, in some cases each task can be decomposed into some subtasks which will then be allocated to different agents, but there are multiple ways for decomposing the task and it is not known prior to task allocation which of the possible decompositions is optimal. Therefore, task decomposing must be performed dynamically during task execution. Note that this type of task is also called complex task and has been systematically studied by Zlot (2006) and some follow-up studies, e.g., (Elmogy *et al.*, 2009; Jiang and Jiang, 2009; Khamis *et al.*, 2011; Landn *et al.*, 2010; Zheng and Koenig, 2009, 2011). We have already introduced this topic in subsection 6.3.2. Secondly, there are cases where the task set may change over time, for example, operators of a multirobot system may submit new tasks or alter or cancel existing tasks. Alternatively, robots may generate new tasks during execution as they observe new information about their surroundings. These type of tasks are sometimes also called online tasks (Dias *et al.*, 2006).

Researches on online task allocation mainly focus on finding out appropriate methodologies for modeling and handling the uncertainties in the task set. A common case is that the task arrivals follow some known probability distributions. Typically, task allocation in this case can be modeled based on Markov Decision Process (MDP) and solved via finding the optimal policies of the MDP (Chapman *et al.*, 2009; Krothapalli and Deshmukh, 2004; Lerman *et al.*, 2006). Krothapalli and Deshmukh (2004) showed that task allocation in computational grids operating in a dynamic and uncertain environment can be modeled as an infinite horizon Markov decision process, with the resource service times and the task arrivals following general probability distributions. Chapman *et al.* (2009) showed that the problem of controlling and managing the allocation of a stochastic set of tasks with varying hard deadlines and processing requirements to a group of agents in real time can be modeled and solved based on a Markov game formulation. Lerman *et al.* (2006) showed that large-size multirobot systems operating in unknown dynamic environments where robots are allowed to change their behavior in response to environmental changes or actions of other robots in order to improve overall system performance can be formalized based on theory of stochastic processes. In this model, robots use a history of local observations of the environment as a basis for making decisions about future actions.

Various optimization goals can be considered for this paradigm. To efficiently compute the optimal policies, Krothapalli and Deshmukh (2004) developed an action elimination procedure for reducing the complexity of computational methods in finding the optimal policy. They also presented a real-time heuristic policy based on certain structural properties of the problem. Chapman *et al.* (2009) argued that in many cases robustness is also a key requirement of the allocation procedure. They focus on the use of a distributed approach in which the individual agents negotiate directly with each other to make the allocations and aim to develop an allocation technique that is, primarily, robust (i.e., decentralized), and secondly, tractable, to ensure solutions can be produced in a timely manner, and finally, efficient, in terms of maximizing the solution quality. To achieve all of these objectives, they pursue a decentralized game-theoretic approach, in which planning is achieved via negotiation between agents.

Market-based approaches can often seamlessly incorporate online tasks by auctioning new tasks as they are introduced by an operator, or as they become available due to the completion of preceding tasks (Dias *et al.*, 2006). Nanjanath and Gini (2006) showed that by using such a paradigm, robustness issues caused by the dynamic environment can be addressed. They presented an sequential auction-based method for the allocation of tasks to a group of robots which operate in a 2D environment, where tasks are locations in the environment that have to be visited by the robots, and unexpected obstacles and other delays may prevent a robot from being able to complete its allocated tasks. While Nanjanath and Gini (2006) focus on addressing robustness issues caused by unreliable robots using sequential auctions, Schoenig and Pagnucco (2010) showed that the optimality with respect to dynamically appearing tasks can also be addressed by market-based approaches. They applied a sequential single-item auction for dynamic task allocation, where not all tasks are known at the start of the auction, and implemented and evaluated two dynamic task allocation schemes providing good compromises between computational complexity and solution quality.

Moreover, bio-inspired approaches are often adopted to solve the optimization problems in dynamic task allocation, for example, artificial immune network (Gao and Wei, 2006), genetic algorithms (Page *et al.*, 2010), and so on. The advantage of these approaches include providing online estimation of resources, dealing with varying resources, dynamically modeling task execution time distributions, and providing an efficient method for scheduling in real-world heterogeneous distributed systems with zero advanced knowledge (Page *et al.*, 2010).

6.7.2 *Task Allocation for Dynamic Agents*

Besides dynamic task sets, dynamic task allocation also includes the problems of allocating tasks to dynamic agents. In this sense, this class also contains the problems studied in task allocation based on online mechanism design, which we have introduced in subsection 6.4.3. Generally, there are two types of dynamic behaviors with respect to the agents.

Firstly, the available agents for executing the tasks may change over time. In this case, a key question is how to properly model the dynamics of the agents and finding effective ways to allocating task in this dynamic environment. Kobayashi *et al.* (2005) studied an open environment with a central manager and a changing number of self interested agents. The central manager is responsible for allocating tasks which arrive from an external source dynamically, at some inter-arrival time between two subsequent occurrences, using a second-price reverse auction as the allocation protocol. The potential number of agents entering the environment between two subsequent auctions is associated with a probability function, and new agents are assumed to enter the environment sequentially, right after an auction, and only if the expected net revenue in this environment is positive for the entering agent. They proposed an efficient algorithm to accurately approximate the agents equilibrium strategies. Macarthur *et al.* (2011) introduced an anytime distributed algorithm for multiagent task allocation problems where the sets of tasks and agents constantly change over time. By using an online pruning procedure that simplifies the problem, and a branch-and-bound technique that reduces the search space, their algorithm can scale to problems with hundreds of tasks and agents.

Secondly, although the agent set is fixed, the properties of the agents such as available resource, ability, cooperation relation, and so on are changing over time. Jung *et al.* (2014) argued that workers may become tired or bored, or begin multitasking, leading to decreased work quality. Alternatively, work quality may improve as a worker's experience with a given task accumulates. To faithfully model such temporal behavior, they presented a time series-based label prediction model for crowd worker's behavioral patterns which may helpfully support better task allocation in crowdsourcing. Pan *et al.* (2016) demonstrated that, for certain types of tasks, workers in crowdsourcing learn from experience and their quality of work may improve over time. They introduced a dynamic hiring mechanism which allocates tasks to workers, not based on their current quality, but based on their learning potential.

6.8   Conclusions

Task allocation has been extensively studied in the past decades with different assumptions and emphases across several subdomains of computer science and technology, and has been formalized as a lot of very different problems, which can be differentiated from each other in the properties of agents, tasks, control models, agent-task relations, optimization objectives, and so on. As a matter of fact, the family of task allocation problems is very large and very complex. This fact makes taxonomy study of task allocation, which aims to summarize and categorize task allocation problems, to be also an important research topic. In this chapter, we have briefly summarized the following six important branches of task allocation research: *taxonomy study on task allocation*, i.e., to study how to summarize and correlate current studies of task allocation and make a common taxonomy; *allocating constrained or complex task*, i.e., to study how to allocate a set of tasks with interdependencies or compound tasks that requires to determine the optimal way of decomposition; *task allocation for rational agents*, i.e., to study how to optimally allocate a set of tasks when the agents are rational players aiming to maximize individual utilities; *task allocation for network systems*, i.e., to deal with the case where the agents are connected by a underlying network, which constraints the interactions of the agents; *distributed task allocation*, i.e., to study how to allocate a set of tasks distributively without the need of a central controller; *dynamic task allocation*, i.e., to deal with the setting in which tasks or agents are dynamic. Intuitively, motivated by current taxonomy study, we categorize the problem space of task allocation based on three axes, agent property, task property, and control model, aiming to develop a global view of task allocation research, which is currently studied in several different subdomains of computer science and artificial intelligence.

We can see a lot of progress has been made on the above directions of task allocation in the past decades. However, there is also a lot of space for further study. For example, taxonomy have been in-depth studied and widely adopted in multirobot task allocation, but a global taxonomy for the general task allocation problems is still missing. Researches on constrained or complex task can further combine current research on operations research, planning, or online algorithms. Task allocation for rational agents is only satisfactorily solved on some special cases, while more general cases, e.g., more general optimization objective, multiple unknown parameters of the agents, are still famous unsolved domains of algorithmic mechanism design. Task allocation in networked systems can incorporate more realistic model of local interaction. Further study on distributed control model of task allocation should develop more perfect approaches for weighing against solution quality and scalability. More models or algorithms for handling uncertainties from planning and scheduling, machine learning, and online algorithms can be tried in dynamic task allocation. More importantly, the task allocation problems introduced in this chapter are intrinsically interrelated with each other. It is of theoretical and practical importance to explore the intersection areas of these problems.

Chapter 7

Automated Negotiation: An Efficient Approach to Interaction Among Agents

Siqi Chen[*,†,§] and Gerhard Weiss[‡]

*College of Computer and Information Science,
Southwest University, Chongqing Shi 400716, China
†Software School, Tianjin University, Tianjin 300072, China
‡Maastricht University, 6200 MD, Maastricht, The Netherlands
§siqi.chen09@gmail.com

Negotiation is any process through which the players on their own try to reach an agreement. It is a task that has a broad spectrum of practical applications to a variety of social, economic, and politic phenomena. When it comes to complicated problems such as negotiations with a large number of issues, finding good agreements is however a tough challenge for human beings, especially in the case that they lack negotiation experience, opponent information, and the available negotiation time is limited. In order to overcome these limitations, there exists considerable interest in automating their negotiation process by means of software agents to assist humans in the decision-making process. Automated negotiation therefore provides people with a realistic alternative solution. This chapter first overviews forms, protocols, and three main approaches of automated negotiation, namely, heuristic, game theoretic, and argumentation approaches. Then, the focus is on the study of complex practical negotiation — multiissue negotiation that runs under real-time constraints and in which the negotiating agents have no prior knowledge about their opponents' preferences and strategies. Finally, two classes of state-of-the-art negotiation agents for complex negotiation are presented, namely, the agents based on regression techniques and the agents based on transfer learning to support its decision-making process during negotiation.

7.1 Introduction

A frequent and important form of interaction in multiagent systems (MAS) is termed automated negotiation. Automated negotiation, as a fundamental and powerful mechanism for managing interaction among autonomous agents, has become a subject of central interest in the MAS community. The main reason behind it is two-fold. First, it can support and facilitate human negotiators in reaching more efficient outcomes by compensating for the limited computational abilities of humans, especially in complicated negotiations, e.g., huge number of rounds, proposals of hundreds of issues, etc. Second, the spectrum of potential real-world applications is very broad and diverse, from electronic commerce, cloud computing, smart grid to supply chain management. In this chapter, we start with the definitions of automated negotiation, and then focus on the study of complex practical negotiation, namely, multiissue negotiation that runs under real-time constraints and in which the negotiating agents have no prior knowledge about their opponents' preferences and strategies. Lastly, two classes of state-of-the-art negotiation agents for complex negotiation are detailed.

7.2 Automated Negotiation

Negotiation can be any process through which the players on their own try to reach an agreement — a task that has many practical applications to a variety of social, economic, and politic phenomena (Muthoo, 1999). Consider the following example: John owns a car that he values at \$25,000 in his opinion, and his friend — Hendric — would like to buy the car at a maximum price of \$50,000. They would therefore have a common interest to have the deal if a contract in which the agreed price lies between \$25,000 and \$50,000 can be made. In addition to those scenarios from everyday life, negotiations may also take place at a larger scale, e.g., diplomatic negotiations over territory dispute between a number of countries.

In many cases, negotiation a time-consuming process, and, moreover, finding good agreements is often a tough challenge for human negotiators, especially when they lack negotiation experience and the negotiation time is limited (Lin and Kraus, 2010). To overcome these limitations, there exist considerable interests in automating their negotiation process by means of autonomous agents to assist humans in the decision-making process. In this context, an agent could be a piece of software (program) that acts on behalf of a certain party and is able to interact autonomously with other agents (or humans) for reaching an agreement on a range of issues that are of conflicting interest to each involved party (Wooldridge and Jennings, 1995). The negotiations operated by autonomous agents are so-called automated negotiations. In contrast with human-based negotiations, automated negotiations enable further complicated negotiation forms because agents have stronger computational power and better memory capability (Williams, 2012). Recent years have witnessed a rapidly growing interest in automated negotiations, mainly due to the broad application range in fields as diverse as electronic commerce and electronic markets (Lau *et al.*, 2008; Ragone *et al.*, 2008), supply chain management (Wang *et al.*, 2009), task and service allocation (Dang and Huhns, 2006), and combinatorial optimization (Duan *et al.*, 2012).

Negotiation can be carried out in a number of ways and forms, such as single-issue or multiissue, bilateral or multilateral negotiations. Several different research trends can be distinguished in automated negotiation models. For example, game-theoretic models consider a negotiation to be a game played by two or more players with a set of actions. Reaching optimal solutions are their main focus under assumptions of full rationality, complete (or partial) information regarding the strategies and preferences of other parties. By contrast, heuristic models aim at searching for solutions to be as close as possible to the optimum, instead of obtaining optimum agreements. Thus, heuristic models assume imperfect knowledge regarding the opponent and the environment, and provide computationally affordable solutions in order to achieve good results. In the following, forms, protocols and main approaches of automated negotiation are overviewed.

7.2.1 *Negotiation Forms*

Due to its diverse applications, automated negotiation could come in a range of forms (Lomuscio *et al.*, 2003; Lopes *et al.*, 2008). The typical (and significant) types are listed as follows:

7.2.1.1 *Single-issue versus multiissue negotiations*

In single-issue negotiations, there is only one issue under consideration for negotiators. To a large extent, it reduces the negotiation complexity since learning opponent in the single-issue case is relatively simple as the loss of one side becomes the gain of the other side. However, the complexity of the negotiation problems increase rapidly as the number of issues grows (i.e., more complex and larger outcome space).

Agent-mediated negotiation with multiple issues places a high demand on agents' capabilities of choosing proper negotiation moves. In such sorts of negotiations, there are likely to be a number of proposals which are at the same utility level from one side's perspective but different from the other side's perspective. The agents, therefore, are expected to find more of those proposals that are mutually beneficial for both parties in order to increase the quality and efficiency of the negotiation.

7.2.1.2 *Bilateral versus multilateral negotiations*

Bilateral negotiation normally refers to the scenario in which two parties negotiate with each other to search for a mutually acceptable agreement between them. The progress of bilateral negotiations is relatively easy to assess because its dominant form of communication can be regarded as offer exchanges between the two parties.

In contrast to bilateral negotiation, multilateral negotiation deals with conflict among multiple parties that sit around the table. The efficiency of multilateral negotiations tends to be too low since multilateral negotiations are time-consuming processes and even need decades to finish (e.g., the negotiation over China's participation in the World Trade Organization). As a matter of fact, multilateral negotiations are far more complex than bilateral ones due to the complicated communication mechanisms, the great variety of interests at stake, the many variables involved in the process and so forth. Therefore, little work has been done in the area of multilateral negotiations, and most research efforts still focus on bilateral negotiations so far.

7.2.1.3 *Sequential versus concurrent negotiations*

When it comes to the negotiation problem that one party needs to negotiate with multiple parties on same items or services, two main options are available for them: sequential or concurrent negotiations (Ponka, 2009). For sequential negotiations, a negotiator negotiates sequentially with others, e.g., all the providers of a service, while in the form of concurrent negotiations, a negotiator negotiates with many or all the providers at the same time. Concurrency has the advantage that participants are allowed to interact with more opponents at a time, thus boosting their learning performance and making high flexibility of decision-making possible (e.g., altering their negotiation tactics in some threads according what they learn from others or seeking for proposals with a better profit given an agreement has been reached in one thread). This feature, on the other hand, gives rise to the management problem of coordinating multiple concurrent negotiating threads, which requires additional computational resources, and gets even worse when concurrent negotiation takes place on a many-to-many basis. In contrast, sequential negotiations are much easier to study and analyze mathematically. The good understanding of sequential negotiations paves the way for building further complicated negotiation forms; therefore in much of the existing work, focus is still on sequential negotiations.

7.2.1.4 *Complete versus incomplete information*

In many cases, information about the opponent is often unknown. Specifically, the opponent's utility function, negotiation strategy and reservation value may be unknown, as revealing such information would lead to exploitation of its behaviors. In a negotiation with incomplete information, it is a real challenge for an agent to know exactly how its actions affect the opponent, as this would depend on information that is unknown. By contrast, if an agent acquires complete information about its opponent, the agent can then determine which outcomes are high performing for both parties. It is clear that the agents in this context are more likely to find a good solution for both sides, and learning opponents (e.g., their utility function, negotiation strategy and the like) is not that important any more.

To summarize, existing literature focuses on automated bilateral sequential multiissue incomplete-information negotiation (e.g., Baarslag *et al.*, 2013; Chen *et al.*, 2013a; Chen and Weiss, 2012, 2014; Lai *et al.*, 2004; Williams *et al.*, 2011), which is also common for real-world negotiation scenarios. Characteristic to this type is that there are two autonomous agents exchanging offers in turn with the goal to agree on a common set of issues such as price, delivery time, quantity and quality, where these issues may be of conflictive importance for the negotiators.

7.2.2 Negotiation Protocol

Negotiation protocol defines the basic rules of the whole negotiation stages, including the types of participants, the negotiation states, the actions that cause negotiation states to change and the actions that participants can make in each state (Jennings *et al.*, 2001). In simple terms, negotiation protocol governs who can be involved in the negotiation as well as how the negotiation should proceed. Some good examples of them include simultaneous offers and alternating offers protocols.

7.2.2.1 Simultaneous offers

Under the simultaneous offers negotiation protocol, the parties on the table make their offers at the same time. The Nash demand game (Nash, 1953) is a single shot simultaneous game, where the two parties simultaneously make a single offer. Consider a concrete negotiation as an example. There are two players negotiating over how to share a single cheesecake. Players p and q make a single offer each, $O_p \in [0,1]$ and $O_q \in [0,1]$, which indicates how much of the pie they wish to eat. If the total amount of pie the players specified in their offers did not exceed one (that is to say $(O_p + O_q < = 1)$, then an agreement is reached between them. Otherwise, if $O_p + O_q > 1$, no agreement is reached, and the negotiation ends in conflict (i.e., negotiation failure). However, it is quite possible that a reached agreement is not efficient, because some of the cake maybe remain unallocated (e.g., $O_p + O_q < 1$). These inefficient outcomes that are not Pareto-optimal[1] fail to maximize the joint expectation of the participating parties.

7.2.2.2 *Alternating offers*

The parties, according to the alternating offers protocol, in turn make offers and counter-offers. This continues until one of the parties accepts the opponent's offer, or alternatively, one of the parties chooses to terminate the negotiation (without agreement), or the negotiation deadline is reached. Compared to the simultaneous offers protocol, the alternating offers protocol is more appropriate for automated negotiation situations where the negotiators are assumed to only have incomplete information most of the time. This is because this protocol could enforce the offers of the two parties compatible and even Pareto-efficient without a costly coordination mechanism. Therefore, automated negotiation strategies tend to be on the basis of the alternating offers protocol.

7.2.3 *Negotiation Approaches*

Although fruitful work on automated negotiation has been contributed to the community of multiagent systems, automated negotiation is generally solved in the following three ways by researchers: heuristic, game-theoretic, and argumentation models.

---

[1]An outcome is considered to be Pareto-optimal, if there are no other outcomes that a single participant could select with no other participants objecting (Wooldridge, 2009).

7.2.3.1 *Heuristic approaches*

Negotiation has traditionally been investigated in game theory (Osborne and Rubinstein, 1994a; Raiffa, 1982), and over the previous years it has also developed into a core topic of MAS (Lopes *et al.*, 2008; Mor *et al.*, 1996; Weiss, 2013). A number of approaches have been proposed that explore the idea to equip an agent with the ability to build a model of its opponent and to use this model for optimizing its negotiation moves (see (Hendrikx, 2011) for a good overview). Modeling the opponent's behavior, however, is practically challenging because negotiators usually do not reveal their true preferences and/or negotiation strategies in order to avoid others from exploiting this information to their advantage (Coehoorn and Jennings, 2004; Raiffa, 1982).

Negotiation approaches employ computationally extensive methods for precise modeling and/or make simplifying assumptions about the negotiation settings (i.e., the environment and the opponent behavior). For example, there are approaches assuming that negotiations are simple single-issue and the opponents have a rather simple (e.g., non-adaptive) behavior. In the following, representative model-based negotiation approaches are overviewed.

Some of the approaches learn opponents by estimating preferences or the reservation value. Faratin *et al.* (2002) proposed a trade-off strategy to increase the chance of getting own proposals accepted without decreasing the own profit. The strategy applied the concept of fuzzy similarity to approximate the preference structure of the opponent and used a hill-climbing technique to explore the space of possible trade-offs for its own offers most likely to be accepted. The effectiveness of this method highly depended on the availability of prior domain knowledge that allowed to determine the similarity of issue values. Coehoorn and Jennings (2004) used Kernel Density Estimation for estimating the issue preferences of an opponent in multiissue negotiations. Their method assumed that the negotiation history was available and that the opponent employed a time-dependent tactic (i.e., the opponent's concession rate depends on the remaining negotiation time, see, e.g., Faratin *et al.* (1998) for details on this kind of tactic). The distance between successive counter-offers was used to calculate the opponent's issue weights and to assist an agent in making trade-offs in negotiation.

Besides, researchers also applied Bayesian learning to automated negotiation. For instance, Zeng and Sycara (1998) used a Bayesian learning representation and updating mechanism to model beliefs about the negotiation environment and the participating agents under a probabilistic framework; more precisely, they aimed at enabling an agent to learn the reservation value of its opponent in single-issue negotiation. Lin *et al.* (2008) used another approach based on Bayesian learning, where the usage of a reasoning model based on a decision-making and belief-update mechanism is proposed to learn the likelihood of an opponent's profile; thereby it is assumed that the set of possible opponent profiles is known as *a priori*. Hindriks and Tykhonov (2008) presented a framework for learning an opponent's preferences by making assumptions about the preference structure and rationality of its bidding process. It is assumed that (i) the opponent starts with optimal bids and then moves toward the bids close to the reservation value, (ii) its target utility can be expressed by a simple linear decreasing function, and (iii) the issue preferences (i.e., issue weights) are obtainable on the basis of the learned weight ranking. Moreover, the basic shape of the issue evaluation functions is restricted to downhill, uphill or triangular. In order to further reduce uncertainty in high-dimensional domains, issue independence is assumed to scale down the otherwise exponentially growing computational complexity. Oshrat *et al.* (2009) developed a negotiating agent for effective multiissue multiattribute negotiations with both human counterparts and automated agents. The successful negotiation behavior of this agent is, to a large extent, grounded in its general opponent modeling component. This component applies a technique known as Kernel Density Estimation to a collected database of past negotiation sessions for the purpose of estimating the probability of an offer to be accepted, the probability of the other party to propose a bid, and the expected averaged utility for the other party. The estimation of these values plays a central role in the agent's decision making. While the agent performs well, the approach taken is not suited for the type of negotiation we are considering (real-time, no prior knowledge, etc.) because opponent modeling is done offline and requires knowledge about previous negotiation traces.

In addition to learning opponents' preferences, other existing approaches try to learn the negotiation strategy and decision model of the opposing negotiator. For instance, Saha *et al.* (2005) applied Chebychev's polynomials to estimate the chance that an opponent accepts an offer relying on the decision history of its opponent. This work dealt with single-issue negotiation, where an opponent's response can only be an accept or a reject. Brzostowski and Kowalczyk (2006) investigated online prediction of future counter-offers on the basis of the past negotiation exchanges by using differentials. They assume that there mainly exist two independent factors that influence the behavior of an opposing agent, namely, time and imitation. The opponent is assumed to apply a weight combination of time- and behavior-dependent tactic.[2] Hou (2004) presented a learning mechanism that employs nonlinear regression to predict the opponent's decision function in a single-issue negotiation setting. Thereby it was assumed that the opponent behavior could only be time-, behavior- or resources-dependent (with decision functions as proposed by Faratin *et al.* (1998)). In the work done by Carbonneau *et al.* (2008) an artificial neural network (ANN) was constructed with three layers that contain 52 neurons to model a negotiation process in a specific domain. The network exploited information about past counter-offers to simulate future counter-offers of opponents. The training process required a very large sample database and therefore became ineffective when applied in online mode.

7.2.3.2 *Game theoretic approaches*

Game theory is the study of strategic behaviors and interaction among autonomous agents. It has been applied to disciplines as diverse as economics, biology, linguistics and computer sciences

---

[2]The concepts of time-dependent and behavior-dependent tactics were introduced in Faratin *et al.* (1998).

(Shoham and Leyton-Brown, 2009). Basically, there are two kinds of game theoretic approaches to negotiation — cooperative and non-cooperative game theory. Cooperative game theory considers games in which it is possible for participants to form coalitions, or in other words the basic unit is a group of agents — so that they together achieve a greater joint utility than they would obtain if they played the game alone. Non-cooperative game theory (Binmore, 1992) instead considers the strategies that can be used by the self-interested negotiation individuals (for maximizing their own utility) and the protocols used in negotiation (for guaranteeing some desirable properties, e.g., guaranteed success, simplicity, individual rationality, stability, etc.).

In a particular agent-based negotiation scenario, game theoretic techniques can help resolve two key problems (Jennings *et al.*, 2001): (1) the design of an appropriate protocol that will govern the interactions between the negotiation participants; and (2) the design of a particular strategy that individual agents can use while negotiating — an agent will aim to use a strategy that maximizes its own individual welfare. While proven to be useful and helpful, there are a number of shortcomings associated with the application of game theory to automated negotiations. One major problem of the theory is related to the strong assumptions about the settings. For instance, game theory assumes that it is possible to characterize an agent's preferences with respect to possible outcomes. However, finding or defining humans' preferences over outcomes are challenging tasks in practice. Furthermore, game theory approaches often assume unbounded computational capability, which means that no computational/time cost needs to be considered in search of mutually accepted solutions within a ranges of outcomes, even when the outcome space is extremely large. In addition, the theory has failed to generate a general model governing rational choice in interdependent situations (Zeng and Sycara, 1997).

Despite these obvious problems, game theory can characterize the solution space of negotiation. One well-known and important concept is Nash Equilibrium, where neither participant can benefit by choosing an alternative action, given that all other participants do not change their action (Nash *et al.*, 1950). For instance, as regards efficiency of a specific solution, Pareto optimality serves as a good indicator — an outcome is said to be Pareto optimal if no negotiation participant can be better off without making at least one other participant worse off. There are also some fairness solution concepts from game theory such as the utilitarian (Myerson, 1981b), Kalai–Smorodinsky (Kalai and Smorodinsky, 1975) considering the fairness of proposed solutions.

7.2.3.3 *Argumentation*

Argumentation-based negotiations provide additional information exchange between negotiators other than the trading of complete (or partial) proposals. This additional information exchanged between agents can be of many different forms, but all of which are arguments which explicitly the reason why an agent makes the argument. As such, a major advantage of exchanging extra information is to enable agents to offer a critique of the received proposal as well as explanations why and where it is not satisfying, rather than simply rejecting the proposal. On the other hand, it is likely for an agent to persuade its negotiation partner with certain arguments to accept a solution by altering its preferences.

In negotiation allowing argumentation, participants benefit from this communication mechanism so that they might reach agreements to which other approaches could not lead. This is however at the expense of significant overheads, due to the reasoning process that the agent has to conduct in order to construct and evaluate the arguments. Moreover, it is also possible for agents to make arguments that are not truthful, which further complicates the negotiation, since the agent needs to evaluate each argument's credibility (Jennings *et al.*, 2001). Lacking of suitable argumentation protocols, building fully functional agents capable of handing practical negotiations are not well addressed so far. Besides, it would be difficult to evaluate the performance of an agent in an environment with varying preferences.

7.3 Characters of Complex Practical Negotiation

The settings of complex negotiations are of relevance to a wide range of practical applications and are also common to many human–human negotiation scenarios. Complex negotiations thus differentiate itself from simple negotiations (e.g., single-issue negotiations, negotiations with prior opponent information, etc.) through the following ways:

7.3.1 *Zero Prior Opponent Knowledge*

In many negotiations, agents are allowed to access some knowledge of the opponents before interacting with them. The information may be vital for decision-making process, including opponents' utility function, negotiation strategy and time constraints. Moreover, in a negotiation with complete information about opponents, an agent is able to calculate which proposals satisfy both sides best. It is therefore much easier for the agents to propose and reach a good agreement. Due to the conflicting nature of negotiations (at least for self-interested participants), opponents however tend to not be willing to reveal this information. In a negotiation where an agent has zero prior knowledge about its opponent, it is hard for it to know exactly which proposals could reach a high level of satisfaction and how the opponent would react to its proposals, as this would depend on opponent information, e.g., preference over negotiation issues, strategies, etc. It thus requires agents to acquire useful opponent information in negotiation via machine learning or other related techniques.

7.3.2 *Continuous-Time Constraints*

For most cases, there is always a deadline by which negotiations should be completed. The constraints of negotiation are a limited amount of continuous-time in many real-life scenarios. The continuous-time constraints refer to constraints based on the amount of physical elapsed time rather than the number of interactions (or rounds) that is studied in previous literature. Negotiators in this context have a real-time deadline to reach an agreement; otherwise, the negotiation ends up as a failure (this deadline is denoted by t_{max} afterward). For instance, in a car sales scenario, the customer needs to make his purchase prior to a specific date. In order to force agreements to be reached in a timely manner, real-time constraints are normally used between the customer and the car dealer.

Due to the effect of the continuous-time constraints, negotiating agents are needed to be more adaptive to opponent strategy (Williams, 2012). That is to say, an agent should not concede too quickly to its opponent, unless the opponent is willing to make a similar compromise. On the other hand, it is also important that an agent does not employ a tough approach (like Boulware strategy (Faratin *et al.*, 2002)), as this could delay the negotiation unnecessarily, causing the value of an agreement to be decreased.

Furthermore, in such contexts, an increased delay grows the likelihood that the agents will fail to reach an agreement. It is therefore a challenge to develop an agent-based strategy that will maximize the agent's gains. Using constraints based on real time makes it more difficult to predict the utility of an offer that an agent might make in the future (other than during the current step). This is because the time at which any future offer will be made by an agent depends on the time that its opponent spends in making its offers. For the same reason, it is also impossible to know how many interactions will occur before the deadline is reached. Therefore, another party influences an agent's utility by delaying the negotiation.

7.3.3 *Discounting Effect and Reservation Value*

The discounting effect is used as an indicator of a cost based on the duration of the negotiation, with the purpose of encouraging agents to complete a negotiation as soon as possible. It has the effect that any particular agreement that is reached at a given time is of higher value than that same agreement at a later time. This causes agents that are slow to reach agreement (e.g., intentionally exploiting opponents) to be punished, as the value of the agreement will be reduced due to the time delay (ANAC, 2012; Williams, 2012).

The reservation utility is the minimum utility level an agent can expect. It also specifies the maximum concession for one side when playing against a non-concessive opponent. Such maximum concession on the other hand restricts the possible exploitation of the opponent, thereby placing higher demand on the flexibility of automated strategy. We next formalize the two factors, beginning with introducing the model of complex negotiation used in existing work.

Let $I = \{a, b\}$ be a pair of negotiating agents, i represent a specific agent ($i \in I$), J be the set of issues under negotiation, and j be a particular issue ($j \in \{1, \ldots, n\}$ where n is the number of issues). The goal of a and b is to establish a contract for a product or service. Thereby a contract consists of a package of issues such as price, quality, and quantity. Each agent has a minimum payoff as the outcome of a negotiation; this is called the reservation value ϑ. Further, w_j^i ($j \in \{1, \ldots, n\}$) denotes the weighting preference which agent i assigns to issue j. The issue weights of an agent i are normalized summing to one (i.e., $\sum_{j=1}^n (w_j^i) = 1$). During a negotiation session, agents a and b act in conflictive roles that are specified by their preference profiles. In order to reach an agreement, they exchange offers (i.e., O) in each round to express their demands. An offer is thereby a vector of values, with one value for each issue. The utility of an offer for agent i is obtained by the utility function defined as

$$U^i(O) = \sum_{j=1}^n (w_j^i \cdot V_j^i(v_{jk})) \qquad (7.1)$$

where v_{jk} is the kth value of the issue j and V_j^i is the evaluation function for agent i, mapping a value of issue j (e.g., v_{jk}) to a real number.

If no agreement can be reached at the end or one side breaks off before deadline, the negotiation then ends up with the disagreement solution (and thus obtain a reservation value ϑ as its payoff). Note that the number of remaining rounds are not known and the outcome of a negotiation depends crucially on the time sensitivity of the agents' negotiation strategies. This holds, in particular, for discounting domains, in which the utility is discounted with time.

The discounting factor $\delta(\delta \in [0,1])$ is used to calculate the discounted utility as follows:

$$D^\delta(U, t) = U \cdot \delta^t \qquad (7.2)$$

where U is the (original) utility and t is the standardized time (i.e., $t \in [0,1]$). As an effect, the longer it takes for agents to ~come to an agreement the lower is the utility they can achieve. Note that a decrease in δ increases the discounting effect.

7.4　State-of-the-Art

The focus of current research is on complex practical negotiations. One of the biggest driving forces behind research into complex negotiations is the broader spectrum of applications over simple negotiations. This emerging research trend in automated negotiations has brought a substantial body of work such as (Chen *et al.*, 2013a, 2013b, 2015; Chen and Weiss, 2013; Hao and Leung, 2012; Williams *et al.*, 2011) that can be roughly classed into two groups. One group is based on regression techniques to learn opponents and adjust its behaviors through the learnt model, while the other is based on transferring knowledge between different tasks and agents creating new sources to aid learning opponents. The remainder of this section illustrates each of them by introducing a typical agent.

7.4.1　*Agents Based on Regression Techniques*

OMAC is a good example (Chen and Weiss, 2013) that models opponents using regression technique. It includes two core stages — opponent modeling and decision-making mechanism. Regression analysis of opponents is done by discrete wavelet transformation (DWT), which is a type of multiresolution wavelet analysis that provides a time–frequency representation of a signal and, based on this, it is capable of capturing time localizations of frequency components. DWT has become increasingly important and popular as an efficient multiscaling tool for exploring features since it can offer with modest computational effort high-quality solutions (with complexity $O(n)$) to non-trivial problems such as feature extraction, noise reduction, function approximation, and signal compression. OMAC employs DWT to extract the main trend of the opponent's concession over time from its previous counter-offers.

In DWT, a time–frequency representation of a signal is obtained through digital filtering techniques, where two sets of functions are utilized: scaling functions using low-pass filters and wavelet functions using high-pass filters. More precisely, a signal is passed through a series of high-pass filters to analyze the high frequencies, and similarly it is passed through a series of low-pass filters to analyze the low frequencies. In so doing, DWT decomposes a signal into two parts, an approximation part and a detailed part. The former is smooth and reveals the basic trend of the original signal, and the latter is rough and in general corresponds to short-term noise from the higher-frequency band.

The decomposition process can be applied recursively as follows:

$$y_{\text{low}}[k] = \sum_n f[n] \cdot h[2k - n]$$

$$y_{\text{high}}[k] = \sum_n f[n] \cdot g[2k - n]$$

(7.3)

with $f[n]$ being the signal, $h[n]$ a halfband high-pass filter, $g[n]$ a halfband low-pass filter, and $y_{\text{low}}[k]$ and $y_{\text{high}}[k]$ the outputs of the low-pass and high-pass filters, respectively. The iterative application of DWT results in different levels of detail of the input signal; in other words, it decomposes the approximation part into a "further smoothed" component and a corresponding detail component. The further smoothed component contains longer period information and provides a more accurate trend of the signal. For instance, f can firstly be decomposed into a rough smooth part (a_1) and a detail part (d_1), and then the resulting part a_1 can be decomposed in finer components, that is, $a_1 = a_2 + d_2$, and so on. This iterative process is captured by the below diagram:

$$f \cdots a_1 \cdots a_2 \cdots a_3 \cdots a_n$$

$$d_1 \quad d_2 \quad d_3 \cdots d_n$$

where $a_1, a_2, \ldots, a_n$ are the approximation parts and $d_1, d_2, \ldots, d_n$ are the detail parts of f.

Given the discrete wavelet function $\psi_{j,k}(t)$ transformed by a mother wavelet $\psi(t)$,

$$\psi_{j,k}(t) = a_0^{-j/2} \psi(a_0^{-j} t - k b_0), \quad j, k \in Z \tag{7.4}$$

DWT corresponds to a mapping from the signal $f(t)$ to coefficients $C_{j,k}$ which are related to particular scales, where these coefficients are defined as follows:

$$C_{j,k} = \int_{-\infty}^{+\infty} f(t) \overline{\psi_{j,k}(t)} dt, \quad j, k \in Z \tag{7.5}$$

The $\psi(t)$ is normally required to be an orthogonal wavelet in practice: the set $\{\psi_{j,k}(t) | j, k \in Z\}$ is then an orthogonal wavelet basis such that the signal $f(t)$ can be reconstructed.

A concrete example of applying DWT in negotiation is given in Fig. 7.1, which shows the curve of the received utilities (i.e., the original signal) in the domain *Airport site selection* when negotiating with the agent IAMhaggler2011. The curve at the top of the figure represents received utilities χ, d_n is the detail component of the nth decomposition layer and a_4 is the approximation on the final layer (i.e., the fourth). This figure clearly shows that a_4 is a pretty good approximation of the main trend of the original signal. As can be seen, the noise/variation represented by those detail components (e.g., d_1–d_4) is irrelevant to its trend.

With DTW, opponent modeling of OMAC is to estimate the utilities of future counter-offers it will receive from its opponent. Opponent modeling is done through a combination of wavelets analysis and cubic smoothing spline. When receiving a new bid from the opponent at the time t_c, the agent records the time stamp t_c and the utility $U(O_{opp})$ this bid has according to the agent's utility function. The maximum utilities in consecutive (equal) time intervals and the corresponding time stamps are used periodically as the basis for predicting the opponent's behavior.

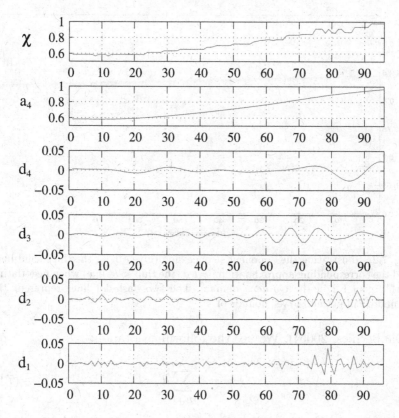

Fig. 7.1. Four-level decomposition of the utilities series χ obtained from IAMhaggler2011 in the negotiation domain *Airport site selection*. The vertical axis shows score and the horizontal axis represents the percentage of time (%). The four-level DWT decomposes the original input χ into five subcomponents including $d_1 \ldots d_4$ (detail parts) and a_4 (approximation part). The resulting components are shown below χ, their frequency increases from d_1 to d_4 (i.e., they are more and more smooth). The final approximation part — a_4 is the long-term trend of χ.

With recursive application of DWT to the signal $f(t)$, the approximation (low-frequency) and detail (high-frequency) components are recovered, respectively. For instance, f can first be decomposed into $a_1 + d_1$ and the resulting part a_1 can then be decomposed in finer components, that is, $a_1 = a_2 + d_2$, and so on. Based upon this recursive process, the signal can be expressed as $f = a_1 + a_2 + \cdots + a_n + d_n$ (further details on wavelets are given in

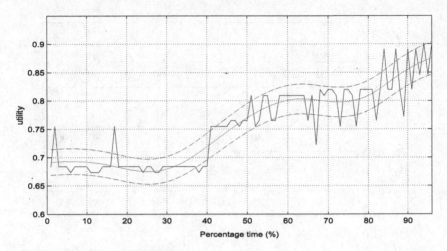

Fig. 7.2. Illustrating the opponent's concession (given by χ, the thick solid line) and the corresponding approximation part v (the thin solid line) when negotiating with Agent_K2 in the *Camera* domain. The two dash-dot lines represent the estimated upper and lower bounds of χ.

(Daubechies, 2006)). We use the following notation:

$$\chi = v + \sum_{n=1}^{\lambda} d_n \tag{7.6}$$

where v represents the approximation component of χ and d_n is n-layer detail part (n is determined by the decomposition level λ). An example can be found in Fig. 7.2 which shows χ and its corresponding approximation part v along with the estimated upper and lower bounds of χ. The two bounds are represented by $v \pm \sigma$, where σ is the standard deviation of the ratio between χ and v.

In order to forecast the opponent's future behavior, cubic smoothing spline is used by OMAC to extend the smooth component v. Cubic spline is widely used as a tool for prediction, see (Yousefi *et al.*, 2005). For equally spaced time series, a cubic spline is a smoothing piecewise function, denoted as the function $\hat{g}(t)$ which minimizes

$$p \sum_{t=1}^{n} w(t)(f(t) - \hat{g}(t))^2 + (1 - p) \int (\hat{g}(u)'')^2 du \tag{7.7}$$

where p is the smoothing parameter controlling the rate of exchange between the residual error described by the sum of squared residuals and local variation represented by the square integral of the second derivative of g and w is the weight vector (for further details, refer to (de Boor, 1978)).

Given the extended version of the smooth part — α — a simple function R, called reserved utility function, is used to realize concession adaptation. This function guarantees the minimum utility at each given time step. This is because the function values are set as the lower bound of our expected utilities. Moreover, in principle, it makes concession over time, thereby taking into account the impact of the discounting factor.

The estimated received utility $E_{ru}(t)$, which indicates the expectation of opponent's future concession, is defined as follows:

$$E_{ru}(t) = D(\alpha(t)(1 + \text{Stdev}(\text{ratio}_{[t_b, t_c]})), t), \quad t \in [t_c, t_s] \qquad (7.8)$$

where $\text{Stdev}(\text{ratio}_{[t_b, t_c]})$ is the standard deviation of ratio between the smooth part v and the original signal χ from the beginning of negotiation(t_b) till now and t_s is the end of α.

Suppose the future expectation the agent has obtained from $E_{ru}(t)$ is optimistic, in other words, there exists an interval $\{T | T \neq \varnothing, T \subseteq [t_c, t_s]\}$, so that

$$E_{ru}(t) \geq D(R(t), t), \quad t \in T \qquad (7.9)$$

OMAC then sets the time $\hat{t}$ at which the optimal estimated utility $\hat{u}$ is reached as:

$$\hat{t} = \underset{t \in T}{\arg\max} \left(E_{ru}(t) - D(R(t), t) \right) \qquad (7.10)$$

and $\hat{u}$ is simply assigned by:

$$\hat{u} = E_{ru}(\hat{t}) \qquad (7.11)$$

7.4.2 *Agents Based on Transfer Learning*

Even though autonomous agents have gained remarkable achievements in various negotiation domains, a significant shortcoming shared by autonomous agents is the omission of knowledge reuse. The problem of opponent modeling is tough due to the lack of enough information about opponents. Knowledge reuse in the form of transfer can serve as a potential solution for such a challenge. Transfer is substantially more complex than simply using the previous history encountered by the agent. Since for instance, the knowledge for a target agent can potentially arrive from a different source agent negotiating in a different domain, a mapping to correctly configure such knowledge is needed in such a case. Transfer is of great value for a target negotiation agent in a new domain. The agent can use this "additional" information to learn about and adapt to new domains more quickly, thus producing more efficient strategies.

Chen *et al.* (2013b) firstly applied this novel transfer learning mechanism to automated negotiation. Specifically, the proposed agent adopts conditional restricted Boltzmann machines (CRBMs) as the basis for opponent modeling, which is in turn used for determining a negotiation strategy. CRBMs, introduced by Taylor and Hinton (2009) are rich probabilistic models used for structured output predictions. CRBMs include three layers: (1) history, (2) hidden, and (3) present layers. These are connected via a three-way weight tensor among them. CRBMs are formalized using an energy function. Given an input data set, these machines learn by fitting the weight tensor such that the energy function is minimized. Although successful in modeling different time series data (Mnih *et al.*, 2012; Taylor and Hinton, 2009), full CRBMs are computationally expensive to learn. Their learning algorithm, contrastive divergence (CD), incurs a complexity of $\mathcal{O}(N^3)$. Therefore, a factored version, the factored conditional restricted Boltzmann machines (FCRBM), has been proposed by Taylor and Hinton (2009). FCRBM factors the three-way weight tensor among the layers, reducing the complexity to $\mathcal{O}(N^2)$. Next the mathematical details will be explained.

Define $\mathcal{V}_{<t} = [v_{<t}^{(i)}, \ldots, v_{<t}^{(n_1)}]$, with n_1 being the number of units in the history layer. Further, define $\mathcal{H}_t = [h_t^{(1)}, \ldots, h_t^{(n_2)}]$, with n_2 being the number of nodes in the hidden layer. Finally, define $\mathcal{V}_t = [v_t^{(1)}, \ldots, v_t^{(n_3)}]$, with n_3 being the number of units in the present layer.

In automated negotiation, the inputs are typically continuous. Therefore, for the history and present layers, a Gaussian distribution, is adopted, with a sigmoidal distribution for the hidden.

The visible and hidden units joint probability distribution is given by

$$p(\mathcal{V}_t, \mathcal{H}_t | \mathcal{V}_{<t}, \mathbf{W}) = \exp\left(\frac{-E(\mathcal{V}_t, \mathcal{H}_t | \mathcal{V}_{<t}, \mathbf{W})}{Z}\right)$$

with the factored energy function determined using

$$E(\mathcal{V}_t, \mathcal{H}_t | \mathcal{V}_{<t}, \mathbf{W}) = -\sum_i \frac{(v_t^{(i)} - a^{(i)})}{\sigma_i^2} - \sum_j h_t^{(j)} b^{(j)}$$
$$- \sum_f \left(\sum_i \mathbf{W}_{if}^{\mathcal{V}_t} \frac{v_t^{(i)}}{\sigma_i} \sum_j \mathbf{W}_{jf}^{\mathcal{H}} h_t^{(j)} \sum_k \mathbf{W}_{kf}^{\mathcal{V}_{<t}} \right)$$

where Z is the potential function, f is the number of factors used for factoring the three-way weight tensor among the layers, and σ_i is the variance of the Gaussian distribution in the history layer. Furthermore, $\mathbf{W}_{if}^{\mathcal{V}_t}$, $\mathbf{W}_{jf}^{\mathcal{H}}$, and $\mathbf{W}_{kf}^{\mathcal{V}_{<t}}$ are the factored tensor weights of the history, hidden, and present layer, respectively. Finally, $a^{(i)}$ and $b^{(j)}$ are the biases of the history and hidden layers, respectively.

Since there are no connections between the nodes of the same layer, inference is done parallel for each of them. The values of the jth hidden unit and the ith visible unit are, respectively, defined as follows:

$$s_{j,t}^{\mathcal{H}} = \sum_f \mathbf{W}_{jf}^{\mathcal{H}} \sum_i \mathbf{W}_{if}^{\mathcal{V}} \frac{v_t^{(i)}}{\sigma_i} \sum_k \mathbf{W}_{kf}^{\mathcal{V}_{<t}} \frac{v_{<t}^{(k)}}{\sigma_k} + b^{(j)}$$

$$s_{i,t}^{\mathcal{V}} = \sum_f \mathbf{W}_{if}^{\mathcal{V}} \sum_j \mathbf{W}_{jf}^{\mathcal{H}} h_t^{(j)} \sum_k \mathbf{W}_{kf}^{\mathcal{V}_{<t}^{(k)}} \frac{v_{<t}^{(k)}}{\sigma_k} + a^{(i)}$$

These are then substituted to determine the activation probabilities of each of the hidden and visible units as

$$p(h_t^{(j)} = 1 | \mathcal{V}_t, \mathcal{V}_{<t}) = \text{sigmoid}(s_{j,t}^{\mathcal{H}})$$
$$p(v_t^{(i)} = x | \mathcal{H}_t, \mathcal{V}_{<t}) = \mathcal{N}(s_{i,t}^{\mathcal{V}}, \sigma_i^2)$$

Learning in the full model means to update the weights when data is available. This is done using persistence contrastive divergence proposed in Mnih *et al.* (2012). The update rules for each of the factored weights are:

$$\Delta \mathbf{W}_{if}^{\mathcal{V}} \propto \sum_t \left(\left\langle v_t^{(i)} \sum_k \mathbf{W}_{kf}^{\mathcal{V}_{<t}} v_{<t}^{(k)} \sum_j \mathbf{W}_{jf}^{\mathcal{H}} h_t^{(j)} \right\rangle_0 \right.$$
$$\left. - \left\langle v_t^{(i)} \sum_k \mathbf{W}_{kf}^{\mathcal{V}_{<t}} v_{<t}^{(k)} \sum_j \mathbf{W}_{jf}^{\mathcal{H}} h_t^{(j)} \right\rangle_K \right)$$

$$\Delta \mathbf{W}_{jf}^{\mathcal{H}} \propto \sum_t \left(\left\langle h_t^{(j)} \sum_k \mathbf{W}_{kf}^{\mathcal{V}_{<t}} v_{<t}^{(k)} \sum_i \mathbf{W}_{if}^{\mathcal{V}} v_t^{(t)} \right\rangle_0 \right.$$
$$\left. - \left\langle h_t^{(j)} \sum_k \mathbf{W}_{kf}^{\mathcal{V}_{<t}} v_{<t}^{(k)} \sum_i \mathbf{W}_{if}^{\mathcal{V}} v_t^{(i)} \right\rangle_K \right)$$

$$\Delta \mathbf{W}_{kf}^{\mathcal{V}_{<t}} \propto \sum_t \left(\left\langle v_{<t}^{(k)} \sum_i \mathbf{W}_{if}^{\mathcal{V}_t} v_t^{(i)} \sum_j \mathbf{W}_{jf}^{\mathcal{H}} h_t^{(j)} \right\rangle_0 \right.$$
$$\left. - \left\langle v_{<t}^{(k)} \sum_i \mathbf{W}_{if}^{\mathcal{V}_t} v_t^{(i)} \sum_j \mathbf{W}_{jf}^{\mathcal{H}} h_t^{(j)} \right\rangle_K \right)$$

$$\Delta a^{(i)} \propto \sum_t (\langle v_t^{(i)} \rangle_0 - \langle v_t^{(i)} \rangle_K)$$
$$\Delta b^{(j)} \propto \sum_t (\langle h_t^{(j)} \rangle_0 - \langle h_t^{(j)} \rangle_K)$$

where $\langle \cdot \rangle_0$ is the data distribution expectation and $\langle \cdot \rangle_K$ is the reconstructed distribution after K-steps sampled through a Gibbs sampler from a Markov chain starting at the original data set.

Upon receiving a new counter-offer from the opponent, the agent records the time stamp t_c, the utility of the latest offer proposed by the agent $u_{\text{own}}^{(l)}$ and the utility, u_{rec}, in accordance with the agent's own utility function. Since the agent may encounter a large amount of data in a single session when operating in real time, the CRBM is trained every short period (interval) to guarantee its robustness and effectiveness in the real-time environments. The number of equal intervals is denoted by ζ.

The agent predicts the optimal utility to receive from the opponent, $u_{\text{rec}}^{\star}$, by training the conditional restricted Boltzmann machine with $(t_c, u_{\text{own}}^{(l)})$ as inputs.

To maximize the received utility from the opponent, the agent should select the optimal offer that maximizes

$$\max_{t, u_{\text{own}}} u_{\text{rec}} = \max_{t, u_{\text{own}}} \mathcal{N}(\mu, \sigma^2) \tag{7.12}$$

$$\text{s.t. } t \leq T_{\max} \tag{7.13}$$

with, $\mu = \sum_{ij} \mathbf{W}_{ij1} v^{(i)} h^{(j)} + c^{(1)}$ being the mean of the Gaussian distribution of the node in the present layer of the Boltzmann machine. In other words, the output of the CRBM on the present layer is the predicted utility, u_{rec}, of the opponent. To solve the maximization problem of Eq. (7.13), the Lagrange multiplier technique is used. Namely, the above problem is transformed to the following:

$$\max_{t, u_{\text{own}}} J(t, u_{\text{own}}) \tag{7.14}$$

where $J(t, u_{\text{own}}) = [\mathcal{N}(\sum_{ij} \mathbf{W}_{ij1} v^{(i)} h^{(j)} + c^{(1)}, \sigma^2) - \lambda(T_{\max} - t)]$. The derivatives of Eq. 7.14 with respect to t and u_{own} are calculated as follows:

$$\frac{\partial}{\partial t} J(t, u_{\text{own}}) = \frac{1}{2\sigma^2} (u_{\text{rec}} - \mu) \mathcal{N}(\mu, \sigma^2) \left[\sum_j \mathbf{W}_{1j1} h^{(j)} \right]$$

$$\frac{\partial}{\partial u_{\text{own}}} J(t, u_{\text{own}}) = \frac{1}{2\sigma^2} (u_{\text{rec}} - \mu) \mathcal{N}(\mu, \sigma^2) \left[\sum_j \mathbf{W}_{2j1} h^{(j)} \right]$$

These derivatives are then used by gradient ascent to maximize Eq. (7.14). The result is a point $\langle t^{\star}, u_{\text{own}}^{\star} \rangle$ that corresponds to $u_{\text{rec}}^{\star}$.

Having obtained $\langle t^{\star}, u_{\text{own}}^{\star} \rangle$, the agent has now to decide on how to concede to $t^{\star}$ from t_c. Williams *et al.* (2011) adopted a linear concession strategy. However, such a scheme potentially omits a lot of relevant information about the opponent's model. To overcome

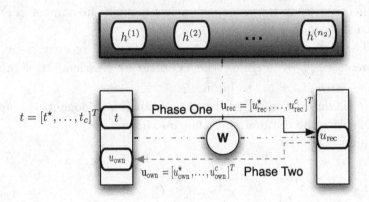

Fig. 7.3. High-level schematic of the overall concession rate determination.

such a shortcoming, this work uses the functional (i.e., learnt by the CRBM) manifold in order to determine the concession rate over time. First, a vector of equally spaced time intervals between $t^\star$ and t_c is created. This vector is passed to the CRBM to predict a vector of relevant u_{rec}'s (i.e., phase one in Fig. 7.3). Having attained u_{rec}, the machine runs backwards, denoted by phase two in Fig. 7.3 (i.e., CD with fixing the present layer and reconstructing on the input) to find the optimal utility to offer at the current time t_c. It is clear that the concession rate follows the manifold created by the factored three-way weight tensor, and thus follows the surface of the learnt function. Adopting the above scheme, the agent potentially reaches $\langle t^\star, u_{own}^\star \rangle$, which are used to attain $u_{rec}^\star$ such that the expected received utility can be maximized.

Given a utility (u_τ) to offer, the agent needs to validate whether the utility of the counter-offer, u_{rec}, is better than u_τ and the (discounted) reservation value, or whether it had been already proposed earlier by the agent. If either of the conditions is met, the agent accepts this counter-offer and an agreement is reached. Otherwise, the proposed method constructs a new offer which has a utility of u_τ. Furthermore, for negotiation efficiency, if u_τ drops below the value of the best counter-offer, the agent chooses that best counter-offer as its next offer, because such a proposal tends to well satisfy the expectation of the opponent, which will then be inclined to accept it.

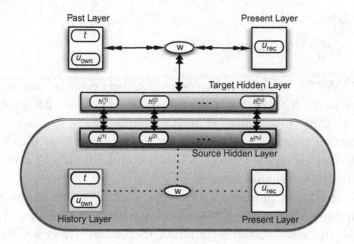

Fig. 7.4. Transfer CRBMs.

Knowledge transfer has been an essential integrated part of different machine learning algorithms. The idea behind transferring knowledge is that target agents when faced by new and unknown domains may benefit from the knowledge gathered by other agents in different source domain(s). Opponent modeling in complex negotiations is a tough challenge due to lack of enough information about the opponent. Transfer learning is a well-suited potential solution for such a problem.

The overall transfer mechanism of *TCRMB* is shown in Fig. 7.4. After learning in a source task, this knowledge is then mapped to the target. The mapping is manifested by the connections between the two tasks' hidden layers (Fig. 7.4). This can be seen as an initialization of the target task machine using the source model. The intuition behind this idea is that even though the two tasks are different, they might have shared features. This relation is discovered using the weight connections between the hidden layers shown in Fig. 7.4. To learn these weight connections, contrastive divergence on the corresponding layers is performed. The weights are learnt such that the reconstruction error between these layers is minimized. It is worth noting that there is no need for the target hidden units to be same as these of the source. The following update rules for learning the weights are used:

$$\mathbf{W}_{ij}^{\mathcal{S}\to\mathcal{T}(\tau+1)} = \mathbf{W}_{ij}^{\mathcal{S}\to\mathcal{T}(\tau)} + \alpha(\langle h_{\mathcal{S}}^{(i)} h_{\mathcal{T}}^{(j)}\rangle_0 - \langle h_{\mathcal{S}}^{(i)} h_{\mathcal{T}}^{(j)}\rangle_K)$$

where, $\mathbf{W}^{\mathcal{S}\to\mathcal{T}}$ is the weight connection between the source negotiation task, $\mathcal{S}$, and the target task $\mathcal{T}$, τ is the iteration step of contrastive divergence, α is the learning rate, $\langle\cdot\rangle$ is the data distribution expectation, and $\langle\cdot\rangle_K$ is the reconstructed distribution after K-steps sampled through a Gibbs sampler from a Markov chain starting at the original data set.

When the weights are attained, an initialization of the target task hidden layer using the source knowledge is available. This is used at the start of the negotiation to propose offers for the new opponent. The scheme in which these propositions occur is the same as explained previously in the no transfer case. When additional target information is gained, it is used to train the target CRBM as in the normal case. It is worth noting, that this proposed transfer method is not restricted to one source task. On the contrary, multiple source tasks are equally applicable. Furthermore, such a transfer scheme is different than just using the history of the agent. The transferred knowledge can arrive from any other task as long as this information is mapped correctly to suit the target. This mapping is performed using the CD algorithm as described above.

The difference between *TCRMB* and *CRMB* lies in the initialization procedure of the three-way weight tensor at the start of the negotiation. In case of *TCRMB*, the initialization is performed using the transferred knowledge from the source task, while in *CRMB* this is done using the standard method, where the weights are sampled from a uniform Gaussian distribution with a mean and standard deviation determined by the designer.

7.5 Conclusion

This chapter begins with introduction of the popular forms, protocols and approaches of automated negotiation. Then state-of-the-art approaches for complex practical negotiation — multiissue negotiation that runs under real-time constraints and in which the negotiating agents have no prior knowledge about their opponents' preferences and strategies — are studied by two fine examples of agents based on regression techniques and agents based on transfer learning. These two types of techniques are used widely in the field of complex negotiation with good performance across a variety of scenarios.

Despite of the successes gained by those agents, two issues still stand out. First, the computational complexity of learning opponents is still not trivial especially when tackling high-dimensional problems. One possible solution is to further reduce computational complexity via adopting more advanced schemes (e.g., sparse techniques). Another issue is how to extend the proposed agents to other more complex negotiation forms like concurrent multilateral negotiations. As far as we are concerned, there is little research focus on the applicability and generality of negotiating agents.

Acknowledgments

This research is supported by National Natural Science Foundation of China (Program number: 61602391) and is also supported by Southwest University and Fundamental Research Funds for the Central Universities (Grant number: SWU115032, XDJK2016C042).

Chapter 8

Norm Emergence in Multiagent Systems

Tianpei Yang[*,‡], Jianye Hao[*,§], Zhaopeng Meng[†,¶] and Zan Wang[*,‖]

*School of Computer Software, Tianjin University,
Tianjin 300350, China*

†*School of Computer Software, Tianjin University,
Tianjin University of Traditional Chinese Medicine,
Tianjin 300193, China*

‡*tpyang@tju.edu.cn*
§*jianye.hao@tju.edu.cn*
¶*mengzp@tju.edu.cn*
‖*wangzan@tju.edu.cn*

In multiagent systems, social norms play an important role in regulating agents' behaviors and facilitating coordination among them and the functioning of agent societies. However, multiple computational agents usually need to coordinate their behaviors to emerge a norm that all agents follow during interactions where multiple norms coexist [Genesereth *et al.* (1988); Young (1996)]. Therefore, how to evolve one of the multiple norms efficiently and consistently in MAS is the key issue. Furthermore, agents may be faced with the high stochasticity of the environment. Agents need to effectively distinguish between the stochasticity of the environment and the exploration of other agents. Moreover, the possible large norm spaces also add an additional dimension of challenge for norm emergence. With the increase of action spaces, most of the existing approaches usually result in very slow norm emergence or even failure to converge. Hence, the emergence of norms in MASs promises to be a productive and challenging research area. In this chapter, we review a certain number of approaches aiming at addressing norm emergence which can be classified into three categories, i.e., top-down approaches, bottom-up approaches and hierarchical approaches. We also review the studies of fixed-strategy agents and discuss the influence of fixed-strategy agents in details.

8.1 Introduction

Multiagent systems (MASs) evolved from distributed artificial intelligence, aiming at addressing large-scale, complex, real-time, and uncertain information problems which usually cannot be solved efficiently from a single agent's perspective. The study of MASs involves the construction of a single-agent technology, such as modeling, reasoning, learning, and planning, and more importantly, coordination techniques for multiple agents to resolve conflicts and reach consensus.

Norms are one of the most effective approaches to address coordination problems in MASs. In human societies, norms (conventions) routinely guide the choice of human behaviors. Conformity to norms reduces social frictions, relieves cognitive loads on humans, and facilitates coordination. Norms (conventions) are ingrained in our social environments and play a pivotal role in all kinds of business, political, social, and personal choices and interactions.

Many distributed systems can be described as MAS, including electronic institutions (Criado *et al.*, 2011b), distributed robotic systems (Morales *et al.*, 2013a), and traffic control systems (Morales *et al.*, 2011). One of the common problems faced by all these systems is how to achieve effective coordination or collaboration to accomplish overall tasks. Therefore, MASs require effective techniques to avoid conflicts and facilitate the interactions of agents, and researches have shown that social norms can serve as a useful mechanism to regulate the behaviors of agents and to facilitate coordination among them (Ågotnes and Wooldridge, 2010; Airiau *et al.*, 2014; Bianchi *et al.*, 2007, Hao *et al.*, 2015, 2016; Hao and Leung, 2013; Kapetanakis and Kudenko, 2005; Mihaylov *et al.*, 2014; Morales *et al.*, 2013b; Mukherjee *et al.*, 2008; Sen and Sen, 2010; Sen and Airiau, 2007; Shoham and Tennenholtz, 1997; Villatoro *et al.*, 2009, 2011, 2013; Yang *et al.*, 2016; Yu *et al.*, 2013a, 2015a, 2016a, 2016b, 2016c, 2016d, Zhang *et al.*, 2009).

Table 8.1. Two-action coordination game.

Agent 1's actions	Agent 2's actions	
	a	b
a	1	−1
b	−1	1

In MAS, òne commonly adopted characterization of a norm is to model it as a consistent equilibrium that all agents follow during interactions where multiple equivalent equilibria coexist (Young, 1996). Multiple computational agents need to coordinate their actions and the interactions can be formulated as stage games with simultaneous moves made by the players (Genesereth *et al.*, 1988). Such stage games often have multiple equilibria, which makes coordination uncertain. One simple example is illustrated in Table 8.1 in which a norm corresponds to a Nash equilibrium where all agents select the same action. This usually can be modeled as a two-player n-action *coordination game*. Therefore, how to evolve one of the two norms ((a, a) and (b, b)) efficiently and consistently in MAS is the key issue in such kinds of games.

Furthermore, agents may be faced with the high stochasticity of the environment. Agents need to effectively distinguish between the stochasticity of the environment and the exploration of other agents. One representative example is shown in Table 8.2, which we call *fully stochastic coordination game with high penalty*. It is not clear how a population of agents can efficiently evolve toward a consistent norm, given the high dynamics from both the environment and other agents. Moreover, the possible large norm spaces also add an additional dimension of challenge for norm emergence. Until now, most of the existing researches (Airiau *et al.*, 2014; Mukherjee *et al.*, 2008; Sen and Sen, 2010; Sen and Airiau, 2007; Villatoro *et al.*, 2009) on norm emergence in MAS only focus on stage games with relatively small sizes. With the increase of action spaces, most of the existing approaches usually result in very slow norm emergence or even failure to converge. Hence, the emergence of norms in MASs promises to be a productive and challenging research area.

Table 8.2. Fully stochastic coordination game with high penalty.

1's payoff 2's payoff		Agent 2's actions		
		a	b	c
Agent 1's actions	a	8/12	−5/5	−20/−40
	b	−5/5	0/14	−5/5
	c	−20/−40	−5/5	8/12

8.2 Norm Emergence Approaches

Norm emergence problem has received wide attention in MASs literature. Shoham and Tennenholtz (1977) firstly investigated the norm emergence problem in agent society based on a simple and natural strategy — the Highest Cumulative Reward (HCR). In this study, they showed that HCR achieved high efficiency on social conventions in a class of games.

There exist two major approaches for addressing norm emergence problem: the top-down (Ågotnes and Wooldridge, 2010; Hao *et al.*, 2016; Morales *et al.*, 2013b) approach and the bottom-up approach (Hao *et al.*, 2017, 2015; Sen and Airiau, 2007; Yu *et al.*, 2013a). The top-down approach investigates how to efficiently synthesize norms for all agents beforehand. The synthesized norms constrain agents' behaviors by forbidding them from performing one or more actions under certain circumstances, thus achieving effective coordination. However, it is usually difficult to synthesize any norm before agent interactions in distributed multiagent interaction environments since there may not exist such a centralized controller and also the optimal norm may vary frequently as the environment dynamically changes. In contrast, the bottom-up approaches focus on investigating how a norm can emerge through repeated local interactions, which promises to be more suitable for such kinds of distributed and dynamic environments.

However, multiagent reinforcement learning (MARL) algorithms are faced with the problem of slow convergence or even failure on convergence, especially in large domains. Hierarchical framework is introduced to facilitate and accelerate norm emergence. A number of works showed efficiency of hierarchical frameworks on norm emergence in MASs (Abdoos *et al.*, 2013; Campos *et al.*, 2011; Yang *et al.*, 2016; Ye *et al.*, 2011; Yu *et al.*, 2015a; Zhang *et al.*, 2009). Next, we will review existing approaches along the above three lines of researches in details.

8.2.1 *Top-Down Approaches*

Normative systems have gained a lot of attention in MASs literature as an effective technique to regulate the behaviors of agents (Criado *et al.*, 2011a; Hoek *et al.*, 2007; Morales *et al.*, 2011; Shoham and Tennenholtz, 1992). A normative system is a set of rules or norms imposed on an MAS to ensure the desirable global properties are fulfilled. Each norm constrains agents' behaviors by forbidding them from performing one or more actions under certain situations. For example, to avoid crowdedness and traffic accidents, cars traveling through crossroads may implement a norm of "cars encountering green traffic lights can move while those which encountered red traffic lights should not be allowed to move". Thus, normative systems are an important approach to achieve effective coordination among agents in MASs.

The paradigm of normative systems in MASs naturally maps to the law system implemented in human societies, which was firstly proposed by Shoham and Tennenholtz (1992, 1995). They initially defined the problem of synthesizing useful social laws in MASs and investigated the required properties and computational complexity of the synthesis process. Specifically, they proposed the definition of useful social laws, which ensure the transitions of each agent between any focal states. They showed the derivation of useful social law in general is an NP-problem and demonstrated that the conditions of deriving useful social laws can be solved in the polynomial time.

Later, Van der Hoek *et al.* (2007) identified three major computational problems of normative systems, i.e., effectiveness, feasibility, and synthesis. They expressed the goal of social laws in the context of an Alternating-time Temporal Logic (ATL)

(Alur *et al.*, 2002) by transforming the social law problem to an ATL model checking problem and investigated the corresponding computational complexities. However, their framework may synthesize unnecessary norms which over-regulate the behaviors of agents. Later, Morales *et al.* (2011, 2013a) pointed out that a desirable normative system should be both effective and necessary. To synthesize both effective and necessary normative systems, they developed a simulation-based mechanism to automatically synthesize normative systems and applied it to the road junction example to illustrate its effectiveness. Their approach, however, does not provide any theoretical guarantee that the synthesized normative system is effective and minimal (i.e., not containing norms that unnecessarily over-regulate the behaviors of agents).

Later, Ågotnes *et al.* (2012) proposed the conception of conservative social laws. They defined effective social laws as making minimal changes to the original system based on the criterion of the distance metric. The core idea of conservative social laws is to model minimal change principle which is inspired from social laws in human societies. Since the laws with the least effect on the habits of people would be easily accepted by the public. The idea of minimal social laws has already been explored; however, most of previous efforts have been focused on exploring the theoretical boundaries of this conception and not on designing practical algorithms for minimal and effective norm synthesis.

Vasconcelos *et al.* (2009) proposed an approach to detect and resolve norm conflicts using first-order constraint solving techniques. Normative conflicts arise when an action is obliged/permitted to perform an action at the same time, which leads to inconsistencies. Their proposed mechanism can automatically detect and resolve norm conflicts including indirect conflicts and conflicts caused by delegation of actions among agents. However, they simply resolve conflicts among norms without identifying the notion of minimal and effective norms. Thus, how to design and synthesize both effective and minimal normative systems is high significance. Degiovanni *et al.* (2014) proposed an approach for goal operationalization, i.e., to automatically compute the required preconditions and required triggering conditions for operations such that the resulting operations establish a set of goals. Their core idea of iteratively refining the operations based on counter-examples is similar to the first phase of the approach proposed by Hao *et al.* (2016).

Recently, Hao *et al.* (2016) proposed a framework to automatically synthesize both minimal and effective normative systems and protocols leveraging lightweight formal methods. They employed the Alloy modeling language and its associated *Alloy Analyzer* (Jackson, 2002), both of which are designed based on lightweight formal methods. Their norm synthesis framework is shown in Fig. 8.1. The process of norm synthesis is divided into two phases. The first is initial norm synthesis phase that synthesizes normative systems involving only the most generalized norms for a minimal number of agents. The second phase is norm refinement that finds both minimal and effective normative system through iterative verification and heuristic search based on the results from the first phase.

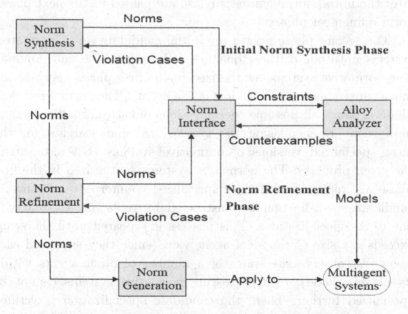

Fig. 8.1. Norm synthesis framework (Hao *et al.*, 2016).

The initial norm synthesis phase generates an initial candidate set of normative systems. The top three components in Fig. 8.1 correspond to initial norm synthesis phase: *Norm Synthesis*, *Norm Interface*, and the *Alloy Analyzer*. It starts by modeling an MAS's behaviors and the desired property in the *Alloy Analyzer*. The analyzer will automatically check whether the MAS satisfies the given desired property, if not, then the analyzer will return a counterexample that shows the sequence of states and actions that causes violation. The *Norm Interface* translates and passes this counterexample to the *Norm Synthesis* component. The *Norm Synthesis* component is responsible for synthesizing normative systems with smallest number of agents which are analyzed for its effectiveness by re-running the *Alloy Analyzer*. At the end of the first phase, the result contains a set of all feasible normative systems that satisfy the given property. The designer enforces either of them on the behaviors of all agent in MAS and reruns *Alloy Analyzer* to generate possible counterexamples separately. This procedure is repeated until no new counterexample is identified. The set of normative systems returned after the initial norm synthesis phase are passed to the next phase: norm refinement phase.

The second phase refines the initial candidate set of normative systems until one of them qualifies as both effective and minimal. The normative systems synthesized in the first phase may put too much constrain on the behaviors of the agents. The norm refinement phase explores all possible specializations of each normative system in a scope by increasing the system size and searches for the most specialized versions of normative systems that also satisfy the given property. The normative systems synthesized in the first phase are replaced with their specialized counterparts. At first, a candidate specialization is constructed by replacing a norm with one of its specializations. This process is repeated until the scope exceeds the size of the local agent view (since they supposed each agent can observe the states of a number of other agents within its neighborhood), or the existing normative systems cannot be specialized further. Then, the candidate specialization is verified against the given property using *Alloy Analyzer*. If the candidate

Norm	Precondition	Actions
n_1	dirNorth(g) $\wedge$ topLeft(h, g) $\wedge$ dirEast(h)	{Move}
n_2	dirNorth(g) $\wedge$ top(h, g)	{Move}
n_3	dirSouth(g) $\wedge$ topLeft(h, g) $\wedge$ dirEast(h)	{Move}
n_4	dirSouth(g) $\wedge$ top(h, g)	{Move}
n_5	dirWest(g) $\wedge$ topLeft(h, g) $\wedge$ dirEast(h)	{Move}
n_6	dirWest(g) $\wedge$ top(h, g)	{Move}
n_7	dirEast(g) $\wedge$ topLeft(h, g) $\wedge$ dirEast(h)	{Move}
n_8	dirEast(g) $\wedge$ top(h, g)	{Move}

Fig. 8.2. The norms generated from the framework proposed by Hao *et al.* (2016).

specialization is effective and cannot be further specialized, then it is stored in as a minimal effective normative system.

They established an automatically minimal effective normative system with a few agents and proved it remains minimally effective with the increase of the number of agents. They also applied their framework to some typical and well-studied multiagent coordination problems used to showcase normative system synthesis methods in MAS, i.e., the road junction example (Kuhn *et al.*, 2010; Morales *et al.*, 2013a, 2011). Figure 8.2 shows the norms generated from their framework. This generated normative systems contains eight norms, where many of them are symmetric, and cover the equivalent scenario under different orientations of the car. They compared their experimental results with the normative using the state-of-the-art framework proposed by Morales *et al.* (2013a) and they proved that their normative system in Fig. 8.2 always guarantees no collision for any general case. They also applied their framework to design a number of population protocols, which are a group of self-regulating network protocols designed for agent-like network nodes. Their designed protocols performed as good as those manually designed and published ones (Angluin *et al.*, 2008; Aspnes and Ruppert, 2010; Kuhn *et al.*, 2010).

8.2.2 *Bottom-Up Approaches*

Until now, significant efforts have been devoted to investigating norm emergence problem from the bottom-up research direction (Airiau *et al.*, 2014; Bianchi *et al.*, 2007; Hao and Leung, 2013; Hao *et al.*, 2015, 2017; Kapetanakis and Kudenko, 2005; Mihaylov *et al.*, 2014; Mukherjee *et al.*, 2008; Sen and Sen, 2010; Sen and Airiau, 2007; Shoham and Tennenholtz, 1997; Villatoro *et al.*, 2009, 2011, 2013; Yu *et al.*, 2013a, 2015a, 2016b–d; Zhang *et al.*, 2009). Sen and Airiau (2007) investigated the norm emergence problem in a population of agents within randomly connected networks where each agent is equipped with certain existing multiagent learning algorithms. They firstly proposed the model of learning *social learning*, where each agent learns from repeated interactions with multiple agents in a given scenario. In this study, the local interaction among each pair of agents is modeled as two-player normal-form games, and a norm corresponds to one consistent Nash equilibrium of the coordination/anti-coordination game.

Later, a number of papers (Airiau *et al.*, 2014; Mukherjee *et al.*, 2008; Sen and Sen, 2010; Villatoro *et al.*, 2009) subsequently extended this work by using more realistic and complex networks (e.g., small-world network and scale-free network) to model the diverse interaction patterns among agents and evaluated the influence of heterogeneous agent systems and space-constrained interactions on norm emergence. Villatoro *et al.* (2009) proposed a reward learning mechanism based on interaction history. In this study, they investigated the influence of different network topologies and the effects of memory of past activities on convention emergence. Later, Villatoro *et al.* (2011, 2013) introduced two rules (i.e., re-wiring links with neighbors and observation) to overcome the suboptimal norm problems. They investigated the influence of Self-Reinforcing Substructure (SRS) in the network on impeding full convergence toward society-wide norms, which usually results in reduced convergence rates.

Additionally, different learning strategies and mechanisms have been proposed to better facilitate norm emergence among agents within different interaction environments (Bianchi *et al.*, 2007; Hao and Leung, 2013; Hao *et al.*, 2015, 2017; Mihaylov *et al.*, 2014; Savarimuthu *et al.*, 2011; Yu *et al.*, 2013a, 2016b, 2016d; Zhang *et al.*, 2009). Savarimuthu *et al.* (2011) recapped the existing mechanisms on the multiagent-based emergence, and investigated the role of three proactive learning methods in accelerating norm emergence. The influence of liars on norm emergence is also considered, and simulation results showed that norm emergence can still be sustained in the presence of liars. Hao and Leung (2013) investigated the problem of coordinating towards optimal joint actions in cooperative games under the social learning framework by introducing two types of learners (IALs and JALs).

Yu *et al.* (2013a) proposed a novel collective learning framework to investigate the influence of agent local collective behaviors on norm emergence in different scenarios and defined two strategies (collective learning-l and collective learning-g) to promote the emergence of norms where agents are allowed to make collective decisions within networked societies. Recently, Yu *et al.* (2016b) proposed an adaptive learning framework for efficient norm emergence. Later, Yu *et al.* (2016d) proposed a novel adaptive learning to facilitate consensus formation among agents in order to establish a consistent social norm in agent society more efficiently.

However, many previous works could not handle the case that norms in practical scenarios may correspond to the forms of Nash equilibria requiring agents to choose different actions. Furthermore, more complicated scenarios involve multiple Nash equilibria, while only some of them correspond to norms which causes high mis-coordination problems. Moreover, it would be more challenging for the norm(s) to evolve if the interaction environment becomes stochastic. To address the above challenges, Hao *et al.* (2015, 2017) proposed two learning strategies under the collective learning framework: collective learning EV-l and collective learning EV-g to address the problem of high mis-coordination cost and stochasticity in complex and dynamic interaction scenarios. The two novel learning strategies are under the networked collective learning framework, which is applicable to a wide variety of scenarios for norm emergence.

Their learning strategies are presented as follows. In each round, each agent is assigned as a row player or a column player randomly and interacts with all of its neighbors determined by the underlying network topology and estimates the best action to interact with each neighbor separately, and then synthesizes a best-response action overall to interact with all neighbors. In collective learning EV-l, the overall best-response action is synthesized based on local exploration, while in collective learning EV-g it is based on global exploration. Besides, each agent's learning strategy incorporates both the optimistic assumption and the relative frequency information of each action based on its experience with its neighbors to overcome the possible side effects of high mis-coordination cost and stochasticity of the environment.

Formally, each agent i holds a Q-value $Q_{i,j}(s,a)$, an EV-value $EV_{i,j}(s,a)$ for each action a under each state $s \in \{\text{Row}, \text{Column}\}$ against each of its neighbors j. Q-value $Q_{i,j}(s,a)$ keeps a record of the action a's past performance against neighbor j. Based on the Frequency Maximum Q-value (FMQ) heuristic (Kapetanakis and Kudenko, 2002a), EV-value $EV_{i,j}(s,a)$ considers not only the optimistic assumption, but also the relative frequency of the maximum reward being received for each action which serves as the basis for making decisions. At the end of each round t, each agent i accordingly updates its Q-values against each neighbor j following Eq. (8.1),

$$Q_{i,j}^{t+1}(s,a) = Q_{i,j}^t(s,a) + \alpha_{i,j}^t(s) \times [R_{i,j}^t(s,a) - Q_{i,j}^t(s,a)] \quad (8.1)$$

Then, each agent i evaluates the relative performance $EV_{i,j}(s,a)$ of each action a against the its neighbor j under the current state s as follows:

$$EV_{i,j}^{t+1}(s,a) = Q_{i,j}^{t+1}(s,a) + c \times f^{t+1}(s,a) \times R_{\max}^t(s,a) \quad (8.2)$$

where,

— $R_{\max}^t(s,a) = \max\{R | \langle a, R \rangle \in P_{i,s}^t\}, s \in \{\text{row}, \text{column}\}$, which is the highest payoff,
— $f^{t+1}(s,a)$ is the frequency of receiving the reward of $R_{\max}^t(s,a)$ until now by choosing action a,
— c is the weighting factor determining the relative importance of $R_{\max}^t(s,a)$.

$f^{t+1}(s,a)$ is obtained based on the combination of the history and the current experience with all neighbors and is computed as follows:

$$f^{t+1}(s,a) = f^t(s,a) + \frac{1}{t+1}[f_s^t(s,a) - f^t(s,a)] \qquad (8.3)$$

At last, each agent i synthesizes a best-response action against each neighbor under the row and column role separately. Each agent i then synthesizes the overall best-response actions under both roles based on the majority voting. It is also necessary for the agents to make additionally random explorations to explore actions with possibly better performance. They presented two ways of making explorations based on ϵ-greedy mechanism: local exploration and global exploration. Local exploration is made before synthesizing the overall best-response action. The collective learning strategy with local exploration is denoted as collective learning EV-l. And global exploration is made after the best-response action has been synthesized and the collective learning strategy with global exploration is denoted as collective learning EV-g.

They evaluated the learning performance of both collective learning EV-l and collective learning EV-g and show that both strategies enable agents to reach consistent norms more efficiently in a wider range of games than previous approaches. Figure 8.3 shows their experimental results of dynamics of the average payoff of agents using different learning strategies. For a three-action coordination game with high penalty (CGHP), which is shown in Table 8.3, they showed that agents can successfully converge to one consistent norm (achieve the average payoff of 10) using both collective learning strategies EV-l and EV-g. However, collective learning-l/g (Yu *et al.*, 2013a) and pairwise learning-c (Sen and Airiau, 2007) failed and only achieved the average payoff of 7.

They also investigated the influence of some system parameters on norm emergence, e.g., the influence of different network topologies, and the influence of the size of neighborhood, population and action space on norm emergence. Furthermore, they conducted a full

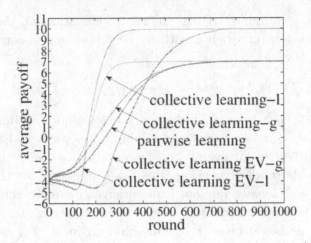

Fig. 8.3. The dynamics of the average payoffs of agents in CGHP under different approaches (Hao *et al.*, 2017).

Table 8.3. Three-action coordination game with high penalty.

1's payoff 2's payoff		Agent 2's actions		
		a	b	c
Agent 1's actions	a	10	0	−30
	b	0	7	0
	c	−30	0	10

analysis on the influence of fixed-strategy agents under collective learning framework, which will be discussed in detail in Section 8.3.

8.2.3 *Hierarchical Approaches*

Hierarchical structures have existed in human societies for a long time. In ancient times, the empire manages the country, and nowadays the government manages citizens and senior governments manage lower level governments. Introducing hierarchical structures to MASs is a meaningful direction to accelerate norm emergence among agents.

Hierarchical learning frameworks, as a promising solution to accelerate coordination among agents, have been studied in different multiagent applications (e.g., package routing (Zhang *et al.*, 2009), traffic control (Abdoos *et al.*, 2013), p2p network (Campos *et al.*, 2011) and smart-grid (Ye *et al.*, 2011)). For example, Zhang *et al.* (2009, 2010) studied the package routing problem and proposed a multilevel organizational structure for automated supervision and a communication protocol for information exchange between higher-level supervising agents and subordinate agents. Simulation shows that the organization-based control framework can significantly increase the overall package routing efficiency than traditional non-hierarchical approaches. Abdoos *et al.* (2013) proposed a multi-layer organizational controlling framework to model large traffic networks to improve the coordination between different car agents and the overall traffic efficiency.

Yu *et al.* (2015a) firstly proposed a hierarchical learning framework to study the norm emergence problem. In this study, they proposed a two-level hierarchical framework. Agents in the lower level interact with each other and report information to their supervisors in the higher level, while agents in the higher level called supervisors to pass down guidance to the lower level. Agents in the lower level follow guidance in policy update. However, their framework is designed for coordination game only, where each agent only needs to coordinate on the same action for norm emergence. However, in realistic interaction scenarios, a norm may correspond to a Nash equilibrium where all agents select different actions. One notable example is considering two drivers arriving at a road intersection from two neighboring roads. To avoid collision, one possible norm is "yield to the left", i.e., waiting for the car on the left-hand side to go through the intersection first. This kind of scenario can naturally be modeled as an *anti-coordination game*, as shown in Table 8.4, which has two different norms, i.e., (a, b) and (b, a).

Most of the previous works only focus on games with relatively small size. This simplification does not accurately reflect the practical interaction scenarios where the action space of agents can be quite large. With the increase of action space, most of the existing

Table 8.4.　Two-action anti-coordination game.

Agent 1's actions	Agent 2's actions	
	a	b
a	-1	1
b	1	-1

approaches usually result in very slow norm emergence or even failure to converge. Furthermore, agents may be faced with the challenge of high mis-coordination cost and stochasticity of the environment. One representative example is shown in Table 8.2, which we call *fully stochastic coordination game with high penalty*. In this game, there exist two optimal Nash equilibria, each of which corresponds to one norm, and one suboptimal Nash equilibrium. Two major challenges coexist in this game: agents are vulnerable to converge to the suboptimal Nash equilibrium due to the high penalty when agents mis-coordinate on the outcomes; agents need to effectively distinguish between the stochasticity of the environment and the exploration of other learners. It is not clear, *a priori*, how a population of agents can efficiently evolve towards a consistent norm, given the large space of possible norms in such challenging environments.

To tackle the above challenges, Yang *et al.* (2016) proposed a Hierarchical Heuristic Learning Strategy (HHLS) under the hierarchical social learning framework to facilitate rapid norm emergence in agent societies. The two-level hierarchical framework is shown in Fig. 8.4. In the hierarchical social learning framework, the agent society is separated into a number of clusters, each of which consist of several subordinate agents. Each cluster's strategies are monitored and guided by one supervisor agent. For each supervisor agent, in each round, it collects the interaction information of its subordinate agents and generates guided instructions in the forms of rules and suggestions for its subordinates. On the other hand, for each subordinate agent, apart from learning from its local interaction, it also adjusts its strategy based on the instructions from its supervisor.

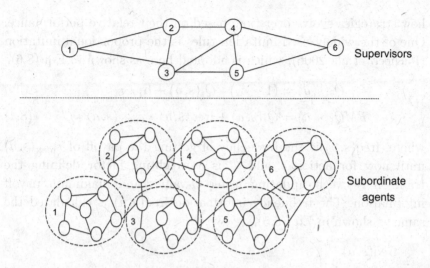

Fig. 8.4. An example of the two-level hierarchical network (Yang *et al.*, 2016).

Formally, each agent i holds a Q-value $Q_i(s,a)$, a FMQ-value $FMQ_i(s,a)$, and an E-value $E_i(s,a)$. In each round, each agent i randomly selects one of its neighbors j determined by the underlying network topology, chooses an action a_i and gets a reward r_i. Then, each agent updates its Q-value following the Eq. (8.4). Additionally, each subordinate agent also updates its Q-values by taking into consideration both the optimistic assumption and the frequency information (Kapetanakis and Kudenko, 2005). Formally we have Eq. (8.4). After that, each agent i reports its information (i.e., the action a_i and reward r_i) to its supervisor.

Each supervisor g also holds a Q-value $Q_g(s,a)$, a FMQ-value $FMQ_g(s,a)$ and an E-value $E_g(s,a)$ for each action a under each state s (row or column player). Each supervisor updates its Q-value, FMQ-value based on all subordinate agents' information accordingly following Eq. (8.4). Then, each supervisor g interacts with a neighboring supervisor m. Specifically, each supervisor communicates with a neighboring supervisor selected based on the neighboring degree between them and imitates the neighbor's strategy. The motivation of imitating peers comes from evolutionary game theory (Weibull, 1997), which provides a powerful methodology to model how strategies evolve over time based on their relative performance. One of the widely used imitation rules is the proportional imitation (Pacheco *et al.*, 2006), which is adopted here as shown in Eq. (8.6).

$$Q_i(s, a) = (1 - \alpha_i) * Q_i(s, a) + \alpha_i * r_i$$

$$FMQ_i(s, a) = Q_i(s, a) + \text{freq}(s, a) * r_{\max}(s, a) * C \qquad (8.4)$$

where, $\text{freq}(s, a)$ is the frequency of getting the payoff of $r_{\max}(s, a)$ until now for action a and C is a weighting factor defining the trade-off between updating using Q-values and maximum payoff information. The frequency information $\text{freq}(s, a)$ is calculated the same as shown in Eq. (8.5).

$$\text{freq}(s, a)$$

$$= \frac{|\{\langle s_m, a_m, r_m \rangle | \langle s_m, a_m, r_m \rangle \in \text{FeedInf}, \\ s_m = s, a_m = a, r_m = r_{\max}(s, a)\}|}{|\{\langle s_m, a_m, r_m \rangle | \langle s_m, a_m, r_m \rangle \in \text{FeedInf}, s_m = s, a_m = a\}|}$$

$$(8.5)$$

where, for each subordinate agent i, its feedback information received by the end of round t as $\text{FeedInf}_i^t = \{\langle s_m, a_m, r_m \rangle | m \in [1, t]\}$.

Each supervisor updates its strategy (denoted as E-value $E_g(s, a)$) for each action a under state s as the average between the FMQ-values of its own and its neighbor m weighted by parameter p following Eq. (8.7). After that, each supervisor generates and issues the instructions (i.e., rules and suggestions) to its subordinate agents. Each supervisor i normalizes the E-values, which serves as the basis for generating instructions for its subordinates. Given a state–action pair $\langle s, a \rangle$, if the normalized E-value $E'(s, a)$ is smaller than a given threshold, it indicates that selecting action a is not a wise choice under state s, thus it is encoded as a rule. The instructions are defined following Eq. (8.8). Finally, each subordinate agent updates its strategy based on the instructions accordingly following Eq. (8.9).

$$p = \frac{1}{1 + e^{-\beta * (FMQ_j(s,a) - FMQ_i(s,a))}} \qquad (8.6)$$

$$E_i(s, a) = (1 - p) * FMQ_i(s, a) + p * FMQ_j(s, a) \qquad (8.7)$$

$$\text{Suggestions} = \{\langle s, a, E_i'(s,a)\rangle \mid a \in A, s \in S\}$$

$$\text{Rules} = \{\langle s, a\rangle | E'(s,a) < \delta\} \tag{8.8}$$

$$E_j(s,a) = FMQ_j(s,a) * (1 + d(s,a) * \rho) \tag{8.9}$$

The main feature of the proposed framework is that through hierarchically supervision among agents, an effective compromising solution can be generated to effectively balance distributed interactions and centralized control towards efficient and robust norm emergence. They evaluated the performance of HHLS under a wide range of games and experimental results show that HHLS can efficiently facilitate rapid emergence of norms compared with the state-of-the-art approaches. For a six-action fully stochastic coordination game with high penalty (FSCGHP), which is shown in Table 8.5, each outcome is associated with two possible payoffs and the agents receive one of them with probability 0.5, which models the uncertainty of the interaction results. The expected payoffs of this game are the same with the CGHP and there also exist four optimal norms marked with bold and two suboptimal norms marked with gray. But it is more complex and difficult to emerge norms due to the stochasticity of the environments.

Figure 8.5 shows the dynamics of the average payoffs of agents in FSCGHP using HHLS compared with the state-of-the-art approaches (Airiau *et al.*, 2014; Yu *et al.*, 2015a). In this challenging game,

Table 8.5. The payoff matrix of six-action fully stochastic coordination game with high penalty.

1's payoff 2's payoff		Agent 2's actions					
		a	b	c	d	e	f
Agent 1's actions	a	**12/8**	5/−5	−20/−40	−20/−40	5/−5	−20/−40
	b	5/−5	14/0	5/−5	5/−5	5/−5	5/−5
	c	−20/−40	5/−5	**12/8**	−20/−40	5/−5	−20/−40
	d	−20/−40	5/−5	−20/−40	**12/8**	5/−5	−20/−40
	e	5/−5	5/−5	5/−5	5/−5	14/0	5/−5
	f	−20/−40	5/−5	−20/−40	−20/−40	5/−5	**12/8**

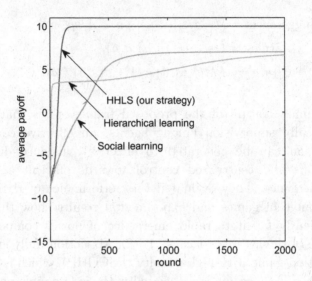

Fig. 8.5. The dynamics of the average payoffs of agents in FSCGHP under different strategies (Yang *et al.*, 2016).

only HHLS enables agents to achieve an average payoff of 10 (agents learn to converge to one optimal). In contrast, the other two state-of-the-art strategies (Airiau *et al.*, 2014; Yu *et al.*, 2015a) converge to one of the suboptimal norms with a slower convergence rate. The superior performance of HHLS is due to the integration of optimistic assumption during strategy update (to overcome miscoordination cost effect) and efficient hierarchical supervision (to accelerate norm emergence speed).

8.3 The Influence of Fixed-Strategy Agents on Norm Emergence

Fixed-strategy agents (FS agents) are those agents who may not have learning abilities and always choose one particular action. It is very common for the existence of multiple norms in multiagent systems. Agents usually do not have preference for any one norm over others in multiagent learning frameworks, e.g., social learning (Airiau *et al.*, 2014; Sen and Airiau, 2007) and collective learning (Hao *et al.*, 2017). The study of fixed-strategy agents has received wide range of attention. Previous works have found that fixed-strategy agents could play a critical role in influencing the direction of norm emergence (Franks *et al.*, 2013; Griffiths and Anand, 2012; Marchant and Griffiths, 2015; Sen and Airiau, 2007).

In this section, we review the studies of fixed-strategy agents in Section 8.3.1 and discuss three kinds of researches about fixed-strategy agents in detail: the influence of fixed-strategy agents on norm adoption in Section 8.3.2, the influence of the placement heuristics of fixed-strategy agents in Section 8.3.3, and the influence of late intervention of fixed-strategy agents in Section 8.3.4.

8.3.1 *Introduction of Fixed-Strategy Agents*

Fixed-strategy agents are those who always select the same action regardless of its efficiency or others' choices. A lot of previous works have shown that inserting small numbers of fixed-strategy agents can significantly influence a larger population of agents when placed in networked social learning framework.

Sen and Airiau (2007) and Airiau *et al.* (2014) firstly investigated the effect of fixed-strategy agents on norm adoption in social learning framework. Their experimental results show that a small number of fixed-strategy agents can influence the whole population towards the particular norm. Griffiths and Anand (2012) proposed a social learning model where each agent has a fixed length FIFO memory which is used to record the most recent actions in history. They investigated the influence of network topologies on norm emergence and demonstrated how fixed-strategy agents can manipulate emergence. They also evaluated strategies for inserting fixed-strategy agents using different placement heuristics in different network topologies.

Franks *et al.* (2013) proposed *Influencer Agents* (IAs) as a mechanism to manipulate norm emergence. IAs are those agents which are inserted with a specific goal of influencing and aiding the emergence of particular norms. They evaluated the influence of IAs in the language coordination domain. Experimental results showed that small proportions of IAs can significantly manipulate the emergence of high-quality norms. Furthermore, they demonstrated the fragility of norm emergence in the presence of malicious or faulty agents that attempted to propagate low quality norms, and verified the importance of network topologies in norm adoption. Franks *et al.* (2014) later proposed an approach for predicting the influence of an agent located at a given location within a network. They evaluated the method in the context of the language coordination domain. They investigated the influence of several placement heuristics of fixed-strategy agents on norm emergence. Experimental results showed that agents can gain much higher influence power using placement heuristics than random placement.

Marchant and Griffiths (2015) investigated norm emergence in dynamic networks and evaluated the effectiveness of placement heuristics of fixed-strategy agents. They defined a Time-To-Live for each agent in a dynamic network and proposed a new heuristic: Life-Degree, combining the age and the degree of each agent. They investigated initial intervention and late intervention of fixed-strategy agents and experimental results showed that inserting small numbers of fixed-strategy agents can influence the whole population converging to the norm adopted by the fixed-strategy agents, even if the population has already established a norm. Further, results showed that a much larger number of fixed-strategy agents is required to affect the whole population using late intervention compared to initial intervention.

Another notable work related with the placement heuristics of fixed-strategy agents is by Tiwari *et al.* (2017) which investigated the effect of leader placement in the context of simulated robotic swarms. Their experimental results showed that the leader placement strategy determines the time it requires for the swarm to converge: leaders who are placed in the middle or periphery of the swarm are better in maneuvering the swarm than those who are placed in front.

Genter *et al.* (2015) investigated the influence of placing fixed-strategy agents in a flock of agents. They aimed to encourage a flock of birds to avoid the dangerous area without significantly disturbing them. They considered inserting fixed-strategy agents to the flock and defined and evaluated several methodologies for placing the influencing agents into the flock. Experimental results showed that the graph approach performed best when using random placement and the grid approach outperformed other approaches when the agents must travel to their desired positions from the position where they are initially placed outside the flock.

This work is extended in Genter and Stone (2016). They investigated where influencing agents should initially be placed within a flock and how influencing agents should join a flock if they can arrive ahead of the flock. Experimental results showed that scaling the area in which fixed-strategy agents are placed to match the area where flocking agents are initially placed fares well. They also showed that hybrid methods combining simple methods with the graph placement approach make a good balance between performance and computation time when initially placing influencing agents into a flock. Lastly, they showed that placement heuristics of the influencing agent that worked well for initial placement also work well if the influencing agents can position themselves within the fight path of an incoming flock.

Recently, Genter and Stone (2017) investigated how robot birds are deployed to influence flocks in nature and proposed a hover approach for robot birds to join and leave a flock. They pointed out the main drawback of the hover approach is that the robot birds are required to hover at desired positions. This drawback may cause hovering problems because robot birds would not be able to hover when they are recognized by natural birds as "one of their own".

8.3.2 *The Influence of Fixed-Strategy Agents on Norm Adoption*

Norms are usually evolved with equal frequency over multiple runs if multiple norms coexist. This is reasonable since all norms in the games (coordination game) are symmetric and the agents do not have any preference over different norms. Airiau *et al.* (2014) investigated the influence of inserting a small number of fixed-strategy agents on norm emergence direction. Their simulations are on a social learning framework. For a two-action stage game which is shown in Table 8.6,

Table 8.6. Stage-games for two agents i and j corresponding to choosing which side of the road to drive on Airiau *et al.* (2014).

	L	R
Agent i		
L	4	−1
R	−1	4
Agent j		
L	2	−10
R	−10	2

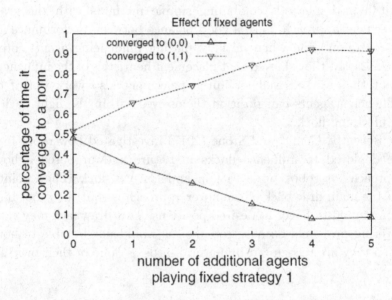

Fig. 8.6. The influence of fixed-strategy agents on norm adoption (Airiau *et al.*, 2014).

the results are showed in Fig. 8.6. Initially both norms ((a, a) and (b, b)) are evolved with equal frequency. As the number of fixed-strategy agents increases, the frequency of evolving norm (b, b) is gradually increased towards 1. The result shows that small amount of agents can have significant influence on the overall population's norm adoption.

Hao *et al.* (2017) investigated the influence of fixed-strategy agents on norm emergence under collective learning framework. Specifically, they considered inserting a number of fixed-strategy agents and investigated the influence of this small amount of fixed-strategy agents on norm adoption of the whole population. Experiments are conducted under the CGHP game (Table 8.3) using collective learning EV-1. Figure 8.7 shows the frequency of the two norms are evolved when the number of fixed-strategy agents increases gradually under the four network topologies. For all four networks, initially both norms ((a, a) and (c, c)) are evolved with equal

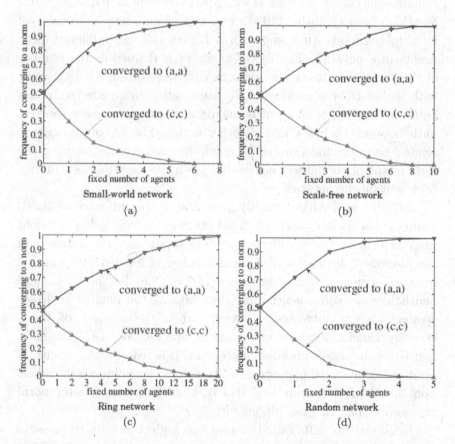

Fig. 8.7. The influence of fixed-strategy agents on norm adoption (Hao *et al.*, 2017).

frequency. As the number of fixed agents increases, the frequency of evolving norm (a, a) gradually increases and reaches 1 eventually.

8.3.3 The Influence of the Placement Heuristics of Fixed-Strategy Agents

There are a lot of metrics used in network topologies to measure and quantify the structural importance of a node, e.g., degree (DC), betweenness (BC), closeness (CC) and eigenvector centrality (EC). Previous work has shown these placement heuristics outperform random placement (Franks *et al.*, 2013; Griffiths and Anand, 2012; Marchant and Griffiths, 2015). The DC of an agent is the size of its neighborhood. An agent with a higher DC can influence more agents in a network. The BC of an agent is defined as the number of the shortest paths in the network that pass through it. An agent with higher BC can propagate the information more effectively in a network. The CC of an agent is calculated using the average shortest path between the node and all other nodes. The EC of an agent is related to the connections of its neighbors and calculated using the eigenvector of the largest eigenvalue given by the adjacency matrix representing the network.

Griffiths and Anand (2012) investigated the influence of fixed-strategy agents for inserting fixed-strategy agents using different placement heuristics in different network topologies. The influence of the placement heuristics of a small number of fixed-strategy agents on norm emergence rate is shown in Fig. 8.8. For all placement heuristics, the convergence time decreases as the number of fixed-strategy agents increases. However, when the number of fixed-strategy agents is greater than 5 or 6, the DC and BC placement heuristics outperform random placement. It is reasonable since fixed-strategy agents with higher centrality values can influence the whole population more quickly and widely, thus facilitating faster norm emergence than random placement.

Franks *et al.* (2014) used several topological metrics to measure the influence of fixed strategy agents on norm emergence. They proposed a methodology for learning the influence of the placement of

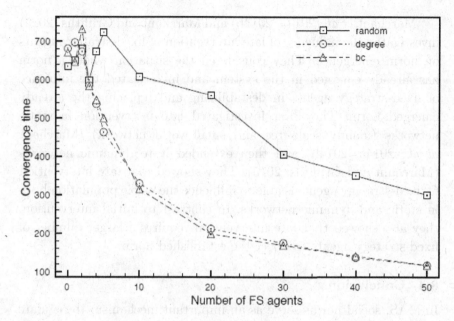

Fig. 8.8. The influence of the placement of fixed-strategy agents on norm emergence rate (Griffiths and Anand, 2012).

agents in a network. Experimental results showed that when placing agents following their methodology, agents can gain much higher influence power than random placement.

8.3.4 *The Influence of Late Intervention of Fixed-Strategy Agents*

There exist two kinds of interventions, e.g., initial intervention and late intervention. Initial intervention refers to inserting fixed-strategy agents at the beginning, while, late intervention means that inserting fixed-strategy agents when a convention has already emerged within the system. In this way, fixed-strategy agents play the role of destabilizing the existed norm. A number of works have shown that using initial intervention of small numbers of fixed-strategy agents has significant influence on the norm emergence direction of a bigger population of agents. There are also some number of works dealing with late intervention which is investigated in the following section.

Marchant *et al.* (2014a, 2014b) and Marchant and Griffiths (2015) investigated the influence of late intervention of fixed-strategy agents on norm emergence. They considered the situation where a norm has already emerged in the system and investigated the influence of fixed-strategy agents in destabilizing and replacing the already emerged norm. They first investigated late intervention in static networks (mainly scale-free and small-world network) (Marchant *et al.*, 2014a, 2014b) and then extended it to dynamic networks (Marchant and Griffiths, 2015). They showed that late intervention of fixed-strategy agents is able to influence the whole population both in static and dynamic networks. In contrast to initial intervention, they also showed that late intervention requires a larger number of fixed-strategy agents to affect the established norm.

8.4 Conclusion

In MAS, social norms serve as an important mechanism to regulate the behaviors of agents and to facilitate coordination among them in MAS. To address the problems of norm emergence among agents, we review a certain number of methods by classifying existing works into three categories, i.e., top-down approaches, bottom-up approaches, and hierarchical approaches. We also review the studies of fixed-strategy agents and discuss the influence of fixed-strategy agents in details.

There are a lot of directions worth further exploring, e.g., how to achieve efficient norm emergence among a larger population of agents under more complex networks, how to design effective multilevel hierarchical structures, and how to design mechanisms for communication and negotiation between agents to facilitate norm emergence.

Chapter 9

Diffusion Convergence in the Collective Interactions of Large-scale Multiagent Systems

Yichuan Jiang*, Yifeng Zhou, Fuhan Yan and Yunpeng Li

*Distributed Intelligence and Social Computing Laboratory,
School of Computer Science and Engineering, Southeast University,
Nanjing 211189, China*

*yjiang@seu.edu.cn

Collective behaviors and phenomena can be often observed in complex systems of nature and human society. How to model and analyze the underlying mechanisms of collective behaviors is a crucial problem for the research of complex systems. The multiagent system (MAS) can be employed as an efficient tool to solve the problem. By designing appropriate large-scale MAS consisting of autonomous agents, the emergence of collective behaviors can be analyzed through the strategy diffusion convergence among the agents having simple interaction rules. In this chapter, we firstly introduce the diffusion convergence of collective behaviors and briefly describe typical diffusion convergence mechanisms in MAS, and then review the researches of diffusion convergence from the following perspectives: (a) the structured and non-structured diffusion convergence according to different organization forms of the systems; (b) the homogeneous and heterogeneous diffusion convergence according to the homogeneity or heterogeneity of influence; and (c) the neighboring and global diffusion convergence according to the difference of sensing scopes of agents.

9.1 Introduction

In current society, complex systems and the relevant complexity problems have become increasingly prominent (Cao *et al.*, 2009; Wang and Lansing, 2004). Complex systems usually consist of many autonomous individuals that can interact with each other. Based on the simple interactions of the autonomous individuals, the systems can emerge with complex behaviors and phenomena. Considering the characteristics of complex systems, researchers proposed the multiagent method to model and analyze complex systems (Jennings, 2001; Liu *et al.*, 2005a).

When the multiagent method is used to model and analyze the complex systems in nature or human society, the systems usually consist of many independent, autonomous, and interactional agents. Each agent has its own behavior rule and decides its behavior strategy based on the contextual information that it perceives. Being inspired from the collective behaviors derived from nature or human society, researchers can design multiagent systems (MAS) to solve large-scale complexity problems, where each agent behaves only by a simple behavior rule (Liu *et al.*, 2005a).

Therefore, collective behavior of multiagents becomes an important issue that cannot be ignored in the researches of complex systems. During the collective interaction processes, agents usually influence each other and then form consentaneous strategy. We refer to this phenomenon as diffusion convergence of collective behavior. In this chapter, we firstly introduce the diffusion convergence phenomenon of multiagents; then we review typical diffusion convergence mechanisms and contrastively analyze their respective characteristics.

The rest of this chapter is organized as follows. In Section 9.2, we introduce the diffusion convergence of collective behavior in MAS and briefly describe typical diffusion convergence mechanisms. In Section 9.3, we contrastively review and analyze structured and non-structured diffusion convergence; in Section 9.4, we contrastively review and analyze homogeneous and heterogeneous diffusion convergence; in Section 9.5, we contrastively review and analyze neighboring and global diffusion convergence. Finally, we conclude this chapter in Section 9.6.

9.2 Diffusion Convergence of Collective Behaviors in MAS

9.2.1 *Collective Behaviors in MAS*

MAS consist of a large number of entities which have different behavior rules and can interact with each other (Jennings, 2001; Liu *et al.*, 2005a; Reynolds, 1987). During the interaction processes, the systems will present some collective behavior characteristics which cannot be observed on single entity (Jennings, 2001; Liu *et al.*, 2005a; Reynolds, 1987). The multiagent collective behavior characteristics can be used in both fundamental research and real applications, such as computer graphics, distributed robotics, traffic control systems and distributed problem solving, etc.

In 1986, Craig Reynolds presented the Boid (Bird-oid) model, which is used to model the flight of flock of birds (Reynolds, 1987). This study (Reynolds, 1987) is regarded as the groundbreaking research of multiagent collective behavior. Since then, multiagent collective behavior has been widely investigated in many fields, such as physics, computer science, and cybernetics, etc. (Jadbabaie *et al.*, 2003; Lin *et al.*, 2004b; Moshtagh and Jadbabaie, 2007; Olfati-Saber *et al.*, 2006, 2007; Tanner and Jadbabaie, 2003; Toner and Tu, 1998; Vicsek *et al.*, 1995). These studies mainly discussed how the agents update their strategies based on their perceived environment and then coordinate with other agents. With time, all the agents will usually achieve harmony with each other and form consentaneous strategy. These studies make a great contribution to modeling the collective behaviors in MAS.

9.2.2 *Diffusion Convergence*

In the researches of multiagent collective behavior, each agent usually has its own strategy. For instance, in a flying flock of birds, the speed and flying direction of a bird can be considered as its strategy. Each bird randomly selects its initial strategy. Then, the birds' strategies will become the same finally. This process is called convergence (Reynolds, 1987). Multiagent behavior convergence can often be observed in natural phenomena. Therefore, the convergence of multiagent collective behavior has attracted the attention of many researchers.

In the researches of multiagent collective behavior, each agent will perceive and imitate the strategies of other agents (this phenomenon is often called diffusion) so that the strategies of all agents will become the same after enough time. The convergence caused by diffusion can be called diffusion convergence (Jiang, 2009). Diffusion convergence can be frequently observed in human society, for example, the custom diffuses in crowd so that the whole society accepts the custom.

According to different mechanisms and forms, diffusion convergence can be classified into many categories. In this chapter, the researches of diffusion convergence are classified as follows: (1) according to the organization forms of MAS: structured diffusion convergence and non-structured diffusion convergence; (2) according to the homogeneity or heterogeneity of influence: homogeneous diffusion convergence and heterogeneous diffusion convergence; and (3) according to the difference of sensing scopes: neighboring diffusion convergence and global diffusion convergence.

We present the summary and analysis of these diffusion convergence mechanisms in the following sections.

9.3 Structured Diffusion Convergence versus Non-structured Diffusion Convergence

During diffusion convergence processes, strategies among a multi-agent group usually diffuse according to a certain mode that will be affected by the organization form of MAS. For any multiagent group, two common organization forms are structured organization form (e.g., artificial social system) and random organization form (e.g., particle swarm system). Under different organization forms, agents usually adopt different strategy diffusion modes. For example, file notices of administrative system are often transmitted along the hierarchical structure. However, social customs in human society usually spread among social groups according to a random diffusion mode. According to the difference of diffusion modes, diffusion convergence can be categorized as structured diffusion convergence and non-structured diffusion convergence.

9.3.1 *Structured Diffusion Convergence*

In realistic MAS, agents may adopt a certain organization structure to cooperate with each other. For example, in artificial social systems that imitate company organizations, hierarchical structures are adopted to coordinate the cooperation among agents. In a MAS with a certain organization structure, each agent occupies a certain interaction position and will influence other agents based on the organization structure. Then, the strategy that an agent adopts is usually related to its interaction position, and the strategy diffusion convergences are constrained by the organization structure of the system. This sort of diffusion convergence can be referred to as structured diffusion convergence. According to the difference of organization structures, structured diffusion convergence can be further categorized as hierarchical diffusion convergence (Bailey, 1982; Hornsby, 2007; Wikle and Bailey, 1997) and equal-rank diffusion convergence (Centola, 2010).

Hierarchical diffusion convergences are commonly seen in social organizations with hierarchical structures. In a hierarchical structure, the interaction position of an agent is not determined by its own features and properties but is determined by its interaction relationships with other agents (Jiang, 2008). The position that is established based on the interaction relationships with other agents is also referred to as social position. During hierarchical diffusion convergences, strategy diffusions generally happen from the agents that hold high social positions to the agents that hold low social positions. For example, in Fig. 9.1, the top-level agent, A_1, finds it easier to diffuse its strategy to the agents in the lower levels and make each agent in the structure have the same strategy.

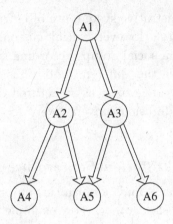

Fig. 9.1. An example of hierarchical structure.

To describe the influence relationship between two agents that occupy different social positions in hierarchical structure, Jiang and Toru adopt the concept of social potential energy (Jiang and Ishida, 2007). In the work of Jiang and Ishida (2007), when describing the diffusion convergence among agents based on potential energy, they assume that each agent is located with a potential field and has its potential energy. Each agent can perceive the potential energies of other agents. Moreover, they propose that the potential energy between two agents is determined by the following aspects: distance between the two agents' social positions (D_{SP}), distance between the two agents' social strategies (D_{SL}), and geographical distance between the two agents' localities (D_{GL}), which is shown as Eq. (9.1).

$$\text{PE}_{ij} = f\left(\frac{\sigma_{SL}D_{SL}(i,j) + \sigma_{SP}D_{SP}(i,j)}{\sigma_{GL}D_{GL}(i,j)}\right) \qquad (9.1)$$

where parameters σ_{SP}, σ_{SL}, σ_{GL} are used to balance the influences of the three kinds of distances on potential energy, and f is a monotonous increasing function.

As Eq. (9.1) shows, the larger the distance between the two agents' social positions or the distance between the two agents' social strategies is, the larger the potential energy between them is.

Meanwhile, the larger the geographical distance between the two agents' localities is, the smaller the potential energy between them is. The potential energy of the whole system is the sum of the potential energies of all agent pairs. Because $PE_{ij} = PE_{ji}$, the potential energy PE of the whole system can be defined as follows:

$$PE = \frac{1}{2} \sum_{i=1}^{n} \sum_{j=1}^{n} PE_{ij} \qquad (9.2)$$

Based on the definition of potential energy of the whole system, along with the strategy diffusion convergence among agents that have different social positions, the potential energy of the whole system will be minimized. Therefore, Jiang and Ishida (2007) use the change of the potential energies among agents to simulate and analyze the diffusion convergence process of agents' strategies in the system.

Equal-rank diffusion convergences are commonly seen in the MAS that are organized based on social networks. A typical equal-rank diffusion convergence in real life is social behavior diffusion convergence (Centola, 2010), where the behavior of a person will be affected by the neighbors in his social network. During equal-rank diffusion convergence processes, each agent has an equal interaction position, and the influence relationship between two neighbor agents is symmetrical. The linear threshold model is a widely used model for equal-rank diffusion convergence, in which an agent is influenced by its active neighbors if the sum of their weights exceeds the threshold of the agent (Jiang and Jiang, 2015). For example, Borodin *et al.* (2010) present a threshold model for a competitive diffusion in social networks. In that model, the agents that have low thresholds can easily adopt others' behavior strategies, and the agents that have high thresholds can adopt others' behavior strategies only after most of the neighbors have adopted that behavior strategy (Jiang and Jiang, 2015).

Centola (2010) investigates the effects of network structure on diffusion convergence of health behavior strategy through artificially structured online communities. In the work, Centola studies the diffusion of a health behavior through a network-embedded population by creating an Internet-based health community, containing 1528 participants recruited from health-interest World Wide Web sites. Based on the empirical experimental results, Centola finds that individual adoption is much more likely when participants receive social reinforcement from multiple neighbors in the social network. The behavior strategy diffuses farther and faster across clustered-lattice networks than across corresponding random networks.

9.3.2 *Non-structured Diffusion Convergence*

Non-structured diffusion convergence can often be seen in real life. A typical non-structured diffusion convergence in real life is collective behavior convergence, where the behavior of the majority of people is easy to diffuse among the crowd and become the consensus of all the people. During non-structured diffusion convergence processes, there is no structural constraint among agents and an agent can influence any other agent in the system.

In the work of Jiang (2009), Jiang proposes a spatial model for collective diffusion convergence. In the model, social distance between agents is measured in an Euclidian space; each strategy s has a certain authority Rank(s) that is determined by not only the number but also the collective positions of its overlaid agents, which is shown as Eq. (9.3).

$$\forall_s \text{Rank}(s) = \sum_{u \in G_s} p_u \qquad (9.3)$$

where G_s represents the overlay group of strategy s, i.e., the set of agents that accept strategy s, and p_u represents the position of agent u. In the model, the strategy that has higher authority is easier to influence the strategies of other agents and to diffuse in the MAS.

During the collective diffusion convergence process, the strategy of an agent is not only affected by the agents in the same overlay group, but also affected by the agents from other overlay groups. Jiang (2009) proposes that the impact force from an overlay group G to an agent a is determined by both their distance and the group's authority. The concrete relation is defined as follows:

$$IF_{G \to a} = f\left(\frac{\sigma_1 \text{Rank}(G)}{\sigma_2 D_{aG}}\right) \tag{9.4}$$

where f is a monotone increasing function, and σ_1 and σ_2 are the parameters to determine the relative importance of the two factors. Additionally, D_{aG} represents the spatial distance between agent a and group G, which is defined as follows:

$$D_{aG} = \frac{1}{|G| \sum_{u \in G} p_u} \sum_{u \in G} (p_u \times d(a, u)) \tag{9.5}$$

where $d(a, u)$ denotes the spatial distance between agent a and agent u. Besides, $|G|$ denotes the number of agents in group G. According to the definition, the nearer the agent is to the group and the higher the group's rank is, the greater the impact force of the group is on agent a.

When a strategy diffuses to an agent, the agent will decide whether to accept the strategy based on its own strategy and position. In a real system, if an agent a is not a member of overlay group G, it may instinctively counter the strategy diffusion influence from group G. Jiang (2009) proposes that the counteracting force from agent a to overlay group G is determined by the following factors: (1) the position of agent a; (2) the average position of overlay group G; and (3) the distance of the strategies between agent a and overlay group G. The concrete relation is defined as follows:

$$CF_{a \to G} = g\left(\alpha_1 \cdot DL_{s_1 s_2} \times \alpha_2 \cdot \left(\frac{p_a}{\frac{1}{|G|} \sum_{u \in G} p_u}\right)\right) \tag{9.6}$$

where s_1 and s_2, respectively, represent the strategies of agent a and overlay group G, and $DL_{s_1 s_2}$ represents the distance between them. In addition, g is a monotone increasing function, and α_1 and α_2 are two weighting parameters. According to the definition, the higher the position of agent a is, the larger the difference between the strategies is, the lower the average position of overlay group G is, and the more likely the agent a resists adopting the social strategy of overlay group G.

When agent a perceives the influence of overlay group G, its strategy is determined by both the impact force of overlay group G and its counteracting force. Jiang (2009) proposes that only if the ratio of the diffusion impact force of a strategy and the counteracting force of an agent exceeds a predefined value, the agent will change its strategy.

In the work of Jiang and Hu (2011), Jiang and Hu present a favor-based model for strategy diffusion in causal multiagent societies. In causal multiagent societies, there are causal–effect relationships among agents. Based on causal-effect relationships, an agent can increase or decrease the utilities of other agents via casual interactions. An agent will favor those who increase (or are likely to increase) its utility, and disfavor those who decrease (or are likely to decrease) its utility. During strategy diffusion processes, an agent is inclined to associate with and imitate the strategies of those who are well-favored. Jiang and Hu (2011) present two methods to model strategy diffusion based on favor: (1) an agent is to imitate the strategy that its most favorable agents hold and (2) an agent is to adjust its strategy based on the strategies of all its favorable agents. Let A be the agent set of the system, s_a be the social strategy of agent a, strategy adoption with consideration of favor is shown as (9.7) or (9.8).

$$\forall a \in A, s_{a_i} = (\gamma_1 \cdot s_{a_i} + \gamma_2 \cdot s_{a_*})|_{a_* = \arg \text{Max}_{a_j \in A}\{f(a_i \to a_j)\}} \quad (9.7)$$

$$\forall a \in A, s_{a_i} = \gamma_1 \cdot s_{a_i} + \gamma_2 \cdot \sum_{a_k \in A_{a_i}^+} \frac{f(a_i \to a_k) \cdot s_{a_k}}{\sum_{a_j \in A_{a_i}^+} f(a_i \to a_j)} \quad (9.8)$$

where γ_1 and γ_2 denote the balance between its own strategy and the diffusion influences, $\gamma_1 + \gamma_2 = 1$. A_{ai}^+ denotes the set of agents that a_i has positive favor level on, and $f(a_i \to a_j)$ denotes the favor level of a_i to a_j.

Based on this diffusion model, Jiang and Hu (2011) find that agents will follow/group with others that have common benefits and keep away from others with whom they have conflicts, which accords with the nature of society.

9.3.3 The Comparison and Analysis of the Two Diffusion Mechanisms

We compare structured diffusion convergence with non-structured diffusion convergence in this subsection, shown in the following:

- **Application scope:** (a) Structured diffusion convergence applies to the MAS with certain organization structure. (b) Non-structured diffusion convergence applies to the MAS with random and unconsolidated organization structure.
- **Percept scope:** (a) In structured diffusion convergence, each agent can only perceive the strategies of its neighbor agents in the system structure and (b) in non-structured diffusion convergence, each agent can perceive the strategy of any agent in the system.
- **Convergence result:** (a) The convergence result in structured diffusion convergence is affected by the system structure and (b) the convergence result in non-structured diffusion convergence is affected by the influences of strategy proponents.

9.4 Homogeneous Diffusion Convergence versus Heterogeneous Diffusion Convergence

The researches of diffusion convergence can be classified into (1) *the homogeneous diffusion convergence* and (2) *the heterogeneous diffusion convergence* in terms of homogeneity or heterogeneity of the MAS. In the following, we concentrate on the homogeneity and heterogeneity of several key factors of the systems that significantly influence the diffusion convergence, such as the agent's social positions, the priority of strategies, the interaction patterns the agents hold, etc.

9.4.1 Homogeneous Diffusion Convergence

The diffusion convergence is homogeneous when the involved factors of agents are the same, namely, the agents may have the same rank during the diffusion process with the same interaction ways. For this reason, all the agents can take approximately the same effect to the strategy diffusion.

The aggregation motion of a flock of birds (Reynolds, 1987) is a familiar phenomenon of the natural world; the birds can fly in approximately the same direction at approximately the same speed. In such systems, each bird can be modeled as an agent, and the flock motion is the aggregation result of the actions of all the agents which have equal ranks, the same strategies, and the same adaptation ways. Hence, the motion of a flock of birds is a typical example of homogeneous diffusion convergence.

Reynolds (1987) proposed the Boid (Bird-oid) model to simulate the motion of a flock of birds. In this model, each bird adjusts its action by only perceiving its neighbors in a local area following three basic rules:

(1) **Collision avoidance:** Each bird avoids collision with nearby flock-mates.
(2) **Direction adjustment:** Each bird attempts to adjust the flying direction as the same with nearby flock-mates.
(3) **Flock centering:** Each bird attempts to stay close to nearby flock-mates.

During the flight, the birds are always changing their strategies following the rules. At the beginning, each bird has a random forward direction from a random position. Then after some time steps, lots of small flocks of birds emerge; in each flock, birds move in approximately the same direction at approximately the same velocity. Small flocks can then merge into a large flock that may be separated into small flocks again in the subsequent flight. The simulation results conducted by the Boid model can coincide with the features of birds' flight in reality. For this reason, this model has been widely used in the design of computer games.

Compared with (Reynolds, 1987), Vicsek *et al.* have proposed a much simpler model in (Vicsek *et al.*, 1995) to simulate the emergence of self-ordered motion of homogeneous particles in biological systems. The only rule of the model is that each particle has a constant absolute velocity but its motion direction is the average of its neighbors within a constant radius with some random perturbation.

$$x_i(t+1) = x_i(t) + v_i(t)\Delta t \tag{9.9}$$

$$\theta(t+1) = \langle \theta(t) \rangle_r + \Delta\theta \tag{9.10}$$

where in (9.9), $x_i(t)$ is the position of a particle at time t, $v_i(t)$ is the velocity of this particle, Δt is the time interval between two updates; in (9.10), $\theta(t)$ represents the direction of the velocity of the particle at time t, $\langle \theta(t) \rangle_r$ is the average direction of velocities of particles within a circle (radius is r), and $\Delta\theta$ is the random perturbation determined with a uniform probability from interval $[-\eta/2, \eta/2]$ where η is a parameter controlling the level of perturbation.

Vicsek *et al.* (1995) concentrate on two basic parameters of this model, the noise involved in the update and the density of the particles in the field. By varying these two parameters, they find two diffusion convergence results:

(1) For small densities and noise, the particles form several groups moving in some random directions, within which the particles move coherently.
(2) For high densities and small noise, the motion of the particles in the field becomes ordered, all the particles move in approximately the same direction.

Furthermore, Jadbabaie *et al.* (2003) provide the theoretical explanation for the observed phenomenon in (Vicsek *et al.*, 1995) based on the algebraic graph theory. Complemented by this work, Vicsek's model can be more practical when it is employed to predict the collective behaviors in reality, such as for the motion of school of fish, flocks of birds, etc.

9.4.2 *Heterogeneous Diffusion Convergence*

When the system is composed of agents with different levels of attributes or different interaction patterns among themselves, e.g., different social positions, strategies, communicate frequencies, etc., the diffusion convergence can be called heterogeneous diffusion convergence. In such a circumstance, agents will make different types and extents of effects on the diffusion convergence.

9.4.2.1 *Agent's heterogeneity*

Jiang and Xia (2009) study the diffusion convergence problem with heterogeneous agents having different social positions and strategy dominance. In the proposed model, different agents have different synchronization capacities defined as:

$$c_i = p_i \cdot \frac{s_i}{\frac{1}{|A|} \sum_{j \in A} s_j} \tag{9.11}$$

where c_i represents the synchronization capacity of a_i, s_i and p_i represent the strategy prominence and the social rank of a_i, and A is the set of agents in the synchronization field. Note that in this model, agents can be influenced by the global context rather than their local neighbors. Supposing that there are two agents a_i and a_j, the synchronization force from a_i to a_j is defined as follows:

$$f(a_i \rightarrow a_j) = \frac{c_i}{c_j} \cdot \frac{1}{d_{ij}^2} = \frac{c_i}{c_j} \cdot \frac{1}{(x_i - x_j)^2 + (y_i - y_j)^2} \tag{9.12}$$

where d_{ij} is the distance between a_i and a_j in the synchronization field.

The most important conclusion presented in this work is that the prominent agents can have higher convergence degrees than other agents in the system.

Gao and Liu (2017) study the diffusion convergence problem of heterogeneous crowds during extreme events. They model the strategy determination of agents as the synchronous update of the interplay of human psychological factors and the external influence on psychological factors:

$$\Delta s_i = \mu F_{SN}(s_i) + (1 - \mu)F_{PM}(s_i) \tag{9.13}$$

where Δs_i is the degree the agent's strategy is influenced; μ is a constant which reflects the relatively dominance of the two types of influences (internal and external). When agents update their strategies according to this mechanism, the perception of risk will largely influence the level of $F_{SN}(s_i)$ and $F_{PM}(s_i)$ and then influence the strategies of the agents. Thus, because of different levels of risk perception of agents in such crowds, the synchronization effects that agents suffered will be different.

In the simulations, Gao and Liu first validate their model by real data obtained from Twitter, then they conduct a series of experiments specially to investigate the effect of heterogeneity of risk perception level of agents on the diffusion convergence (namely the emergence of collective behaviors) in the extreme events. They find that (1) a group of agents with relatively higher risk perception level will trigger other agents' emotion intensity, and then will further motivate the collective behavior and (2) agents closer to this group will be influenced to a greater extent. In fact, these conclusions coincide with the results described by (Jiang and Xia, 2009) to some extent.

9.4.2.2 *Interaction's heterogeneity*

Besides the heterogeneity of agents, the diffusion convergence considering the heterogeneity of interactions among agents has also been studied.

Rahmandad and Sterman (2008) investigate the influence of heterogeneity of interactions among agents and the contact frequency on the diffusion. In this work, the SEIR model is used to model the diffusion among agents; all agents are in one of four states: susceptible (S), exposed (E), infected (I), and recovered (R); these compartments are updated by the expected value of the sum of probabilistic rates for each individual. The agent-based idea is also combined into the SEIR model to reflect the heterogeneity of contact frequency among agents. The contact frequency between two agents depends on two factors: (1) how often the link between them has been used on average and (2) contacts are limited by time constraints (which indicates two aspects: (a) the frequency a link is used may be constant and (b) the frequency a link is used is inversely proportional to the number of links).

With respect to the heterogeneity of contact frequency, Rahmandad and Sterman (2008) mainly find that heterogeneity causes earlier mean peak time which denotes a higher state transition speed. The potential reason is that the link with higher contact frequency will have larger positive effect on the diffusion among agents.

9.4.3 *The Comparison and Analysis of the Two Diffusion Mechanisms*

We present distinguishing features of each type of diffusion convergence from the following aspects:

- **Convergence capacity of agents:** (a) In homogeneous diffusion convergence, agents have the same convergence capacity, since the involved agents have approximately the same attributes, e.g., the equal social rank, the same interaction ways, etc. and (b) inversely, agents have different convergence capacities in heterogeneous diffusion convergence because of their different levels of attributes.
- **Convergence criterion:** (a) The convergence criterion in homogeneous diffusion convergence is that the strategies of agents do not collide with each other and (b) The convergence criterion in heterogeneous diffusion convergence is that the convergence degree of the agents' strategies keeps stable.
- **Convergence result:** (a) The convergence result in homogeneous diffusion convergence is often equal to the average strategy held by agents in the system and (b) The convergence result in heterogeneous diffusion convergence is mainly determined by the strategies held by dominant agents in the system.

9.5 Neighboring Diffusion Convergence versus Global Diffusion Convergence

Based on the sensing scopes of agents in the diffusion convergence process, the studies of diffusion convergence can be categorized as neighboring diffusion convergence and global diffusion convergence.

9.5.1 *Neighboring Diffusion Convergence*

In the neighboring diffusion convergence process, each agent can just perceive the strategies of its direct neighboring agents and then update its own strategy based on the perceived information. Although the strategy of an agent will just be influenced by the neighbors, the influence will diffuse in the MAS based on repeated interactions. Therefore, with time MAS, the convergence will come true in the whole MAS.

In networked MAS, the neighbors of each agent are usually fixed. Each agent can just interact with its direct neighbors in networks (Das *et al.*, 2014). For instance, average model (Das *et al.*, 2014) is used to study how convergence is formed when agents' opinions or strategies are updated using the average strategy of the neighboring agents in a fixed network. Besides, the edges in the network have weights which represent the relationship strengths among agents. In every time step, each agent updates its opinion or strategy based on the weighted average strategy of its neighboring agents. In detail, there are n agents located in an undirected network $G(V, E)$. The link $(i, j) \in E$ has weight $w_{ij} = w_{ji}$. Let $N(i)$ denote the set of agents who have an edge connecting to agent a_i. Let s_i denote the strategy of agent a_i. The rule of strategy update in this model is as follows (Das *et al.*, 2014):

$$s_i(t+1) = \frac{w_{ii}s_i(t) + \sum_{a_j \in N(i)} w_{ij}s_j(t)}{w_{ii} + \sum_{a_j \in N(i)} w_{ij}} \qquad (9.14)$$

Here, w_{ii} is the weight that agent a_i insists its own strategy.

On the other hand, some previous studies (Jadbabaie *et al.*, 2003; Lin *et al.*, 2004b; Vicsek *et al.*, 1995) present a discrete time model in which the neighbors are not fixed. In this model, n autonomous agents move in a plane at same speed. However, the initial moving directions of these agents are various. Each agent may change its moving direction depending on its own moving direction and the average of the moving directions of its neighboring agents. Agent's neighbors at time t are those agents which are either in or on a circle of prespecified radius r centered at agent's current position (Jadbabaie *et al.*, 2003). Although the neighboring agents of an agent change with time, all the agents will move in the same direction after enough time. In detail, let s_i denote the strategy of agent a_i. Then, based on discrete time model, strategy s_i will change with discrete time, such as shown in Eq. (9.15) (Jadbabaie *et al.*, 2003). The left part of Eq. (9.15) is the strategy of agent a_i at time $t+1$. The right part is the average strategy of agent a_i and agent a_i's neighboring agents at time t.

$$s_i(t+1) = \frac{1}{1+|N_i(t)|} \left[s_i(t) + \sum_{a_j \in N_i(t)} s_j(t) \right] \qquad (9.15)$$

Here, $N_i(t)$ represents the set of agents which are the neighbors of agent a_i at time t. $N_i(t)$ can be represented by

$$N_i(t) = \{a_j | d(a_i, a_j) \le r\} \qquad (9.16)$$

In Eq. (9.16), $d(a_i, a_j)$ denotes the physical distance between agents a_i and a_j. The relationships among agents which can perceive each other are symmetrical. In other words, $a_j \in N_i(t) \Rightarrow a_i \in N_j(t)$. Based on Eqs. (9.15) and (9.16), it can be known that each agent changes its strategy depending on the average strategy of its neighboring agents.

Then, the strategy diffuses based on repeated interactions so that convergence comes true in the global system.

9.5.2 *Global Diffusion Convergence*

In some real cases of diffusion convergence, each agent will not only be influenced by its neighboring agents but may also be influenced by the agents which are not its neighbors but sometimes interact with it. Therefore, the concept of global diffusion convergence is presented.

The study by Jiang and Xia (2009) adopts the concept of variable sensing scope which means that the sensing scope of an agent may vary with time and environment change. Although the strategy of an agent will just be influenced by the neighbors, variable sensing scope may lead to global influence. The model in (Jiang and Xia, 2009) defines the sensing scope using a simple method that makes the sensing radius variable. Each agent also updates its strategy based on Eqs. (9.15) and (9.16). The main difference is that, the radius will change, and $r \in (0, \infty)$. If $r \to \infty$, each agent can perceive the strategies of all agents in the system.

Considering hybrid structure of MAS, the study by (Jiang et al., 2007) presents a model which combines neighboring diffusion convergence and global diffusion convergence. In this hybrid model, each agent may be influenced in two ways: (1) neighboring diffusion force: the influence is from neighboring agents and (2) social force: the influence is from the agents in global system. Obviously, if only the first influence is considered, global differentiation may appear. Conversely, if only the second influence is considered, the convergence in neighborhood is ignored. The study by Jiang *et al.* (2007) also presents simulation experiments of the diffusion models. The experimental data show that if an agent prefers to adopt the average strategy of its neighboring agents, convergence will come true more effectively. Besides, the local convergence and global convergence are compatible. The study by Jiang *et al.* (2007) proposes a rule of changing strategy in diffusion convergence process, considering a balance between local influence and global influence, such as shown in Eq. (9.17). In detail, s_i^L denotes the weighted average strategy of agent a_i's neighboring agents. Besides, s_i^S represents the weighted average strategy of the agents which can influence a_i in the global structure.

$$s_i(t+1) = \lambda_L s_i^L(t+1) + \lambda_S s_i^S(t+1)$$

$$= \lambda_L \left[\alpha s(t) + \beta \frac{1}{|N_i(t)|} \sum_{a_j \in N_i(t)} s_j(t) \right] + \lambda_S s_i^S(t+1) \quad (9.17)$$

Obviously, Eq. (9.17) synthetically considers the influence in two aspects. In this equation, the four parameters $(\alpha, \beta, \lambda_L, \lambda_S)$ can be used to determine the tendency in diffusion convergence process: (1) $\alpha + \beta = 1$: the two parameters balance the influence of an agent's own strategy and the average strategy of its neighboring agents. If $\alpha > \beta$, an agent prefers to insist its own strategy. If $\alpha < \beta$, an agent prefers to adopt the average strategy of its neighboring agents. (2) $\lambda_L + \lambda_S = 1$: the two parameters balance the influence from neighborhood and the influence from global social structure. If $\lambda_L > \lambda_S$, an agent prefers to accept the influence from neighborhood. If $\lambda_L < \lambda_S$, an agent prefers to accept the influence from global social structure.

Some studies (e.g., Yan *et al.* (2016)) also discuss the global influence in networked MAS. As each agent can just interact with its direction neighbors in the network, the global influence usually diffuses based on the information which contains opinion or strategy. For instance, the study by Yan *et al.* (2016) discusses the case that each agent can transmit information itself and its own comments. The information itself can be considered as the initial transmitter's opinion or strategy. Therefore, when an agent receives the information, it may be influenced both by its neighbors and the initial transmitter. If there are many diffusion processes in the network, multiple initial transmitters can cause global influence. In detail, $O(i, t)$ denotes agent a_i' opinion at time t. Besides, O_0 denotes the opinion that is contained in the information. In other word, O_0 is the opinion of the initial transmitter. Agent a_i can influence agent a_j according to the probability:

$$p_{i,j} = \begin{cases} p_0 e^{-|O_0 - O(j,t)|}, & \text{agent } a_i \text{ is initial transmitter} \\ p_0 e^{-\frac{1}{2}|O_0 - O(j,t)| - \frac{1}{2}|O(i,t) - O(j,t)|}, & \text{otherwise} \end{cases} \qquad (9.18)$$

Here, p_0 is a basic probability. $p_{i,j}$ is the modification of p_0. Then, A_i denotes the set of opinions which can influence $O(i, t)$, and A_i is ϕ at the beginning of each time step. If agent a_i's neighbor agent a_j has transmitted the information and $|O(i,t) - O(j,t)| < \varepsilon_i$, $O(j,t)$ will be added to A_i. Besides, if $|O(i,t) - O_0| < \varepsilon_i$, O_0 will also be added to A_i. Then,

$$O(i, t+1) = \frac{O(i,t) + \sum_{O(j,t) \in A} O(j,t)}{1 + |A_i|} \qquad (9.19)$$

Here, ε_i is a confidence bound.

9.5.3 The Comparison and Analysis of the Two Diffusion Mechanisms

We compare neighboring diffusion convergence and global diffusion convergence in this subsection, shown in the following.

- **Convergence capacity of agents:** (a) In neighboring diffusion convergence, each agent can just perceive its neighboring agents and (b) in global diffusion convergence, each agent can perceive both its neighboring agents and the global influence in some ways.
- **Convergence criterion:** (a) In neighboring diffusion convergence, each agent updates the strategy based on the strategy of its neighboring agents and (b) in global diffusion convergence, each agent updates the strategy based on both neighboring influence and global influence.
- **Convergence result:** (a) In neighboring diffusion convergence, global convergence comes true based on repeated neighboring interactions and (b) in global diffusion convergence, global convergence and neighboring convergence may simultaneously come true.

9.6 Conclusion

Multiagent simulation is an efficient tool in the researches of complex systems. In the researches of complex systems, the diffusion convergence of collective behaviors has been widely discussed. In this chapter, we firstly introduce the diffusion convergence phenomenon of multiagents; then we review typical diffusion convergence mechanisms from three perspectives and analyze their respective characteristics. The researches of diffusion convergence can be classified as follows: (1) according to the organization forms of MAS: structured diffusion convergence and non-structured diffusion convergence; (2) according to the homogeneity or heterogeneity of influence: homogeneous diffusion convergence and heterogeneous diffusion convergence and (3) according to the difference of sensing scope: neighboring diffusion convergence and global diffusion convergence.

Chapter 10

Incorporating Inference into Online Planning in Multiagent Settings

Yingke Chen*,¶, Prashant Doshi†, Jing Tang‡ and Yinghui Pan§

*Sichuan University, Chengdu 610064, China
† University of Georgia, Atlanta, GA 30302, US
‡ Teesside University, Middlesbrough TS1 3BX, UK
§ Jiangxi University of Finance and Economics,
Nanchang 330000, China
¶ yke.chen@gmail.com

In non-cooperative multiagent settings, methods for planning require modeling other agents' possible behaviors. However, the space of these models — whether these are policy trees, finite-state controllers, or intentional models — is very large and thus arbitrarily bounded. This may *exclude* the true model or the optimal model. In this chapter, we present a novel iterative algorithm for *online* planning that considers a limited model space, updates it dynamically using data from interactions, and provides a provable and probabilistic bound on the approximation error. We ground this approach in the context of graphical models for planning in partially observable multiagent settings — interactive dynamic influence diagrams. We empirically demonstrate that the limited model space facilitates fast solutions and that the true model often enters the limited model space.

10.1 Introduction

Sequential decision making, also known as planning, is an essential task for intelligent agents. Such agents deliberate and take actions along time to either proactively pursue their goals or response to stimulus. The situated environment evolves depending on the performed actions, which always bring a cost or utility to the agent. The agent shall be capable of choosing optimally among possible actions along time. In general, the sequential decision-making problem, after considering uncertainties from different aspects, is a hard problem in terms of computational complexity (Bernstein *et al.*, 2002). As a principled formalism, partially observable Markov decision processes (POMDPs) can unify three types of uncertainties into an enumerative representation: the stochastic dynamics of the situated physical environment, partial observability of the true state of the environment, and the non-deterministic effects of performed actions. After comprehensive consideration, its solution, named as a policy, prescribes the agent's actions given specific observation sequences along time. The aim of solving sequential multiagent decision-making problems is the build policies which maximize (or minimize) the expected utilities (or costs) of agents' behavior.

Besides, agents (named as subject agents) need to cope with other agents (named as object agents) who share the same environment. Those other agents can either be cooperative when they are going to solve a common task together or non-cooperative when they are adversaries. In a cooperative setting, a team of agents seek an optimal joint policy given their common knowledge. Assuming agents receive common reward based on their joint action, the resulting joint policy is optimal for all agents. In contrast, in a non-cooperative setting, as each agent has its own intentions and interests which could be conflict with each other, agents are pursuing their individual optimal from their own perspectives. More importantly, subject agents shall notice that the object agents also deliberate over the environment and other agents for their own sake.

Approaches for planning and plan recognition in cooperative and non-cooperative multiagent settings (Chandrasekaran *et al.*, 2014; Gmytrasiewicz and Doshi, 2005) often model other agents' possible behaviors. These models could be policy trees (Seuken and Zilberstein, 2007), finite-state controllers (Marecki *et al.*, 2008), or intentional models with beliefs, preferences and capabilities (Doshi *et al.*, 2009). Because the space of such models is very

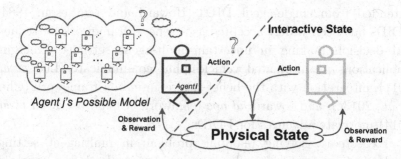

Fig. 10.1. Agent i deliberates over the physical world and another agent j with unknown types. Agent i's decision making has taken agent j's possible decision models into account.

large — theoretically, it is countably infinite — a small subset of models is typically handpicked or arbitrarily selected. However, this precludes any guarantees that the true model or the policy tree that is part of an optimal joint plan is included.

Observing that the true behaviors of other agents are revealed only when the agents interact, we target the following problem setting in this chapter: a subject agent repeatedly interacts with another agent whose behavior (guided by a plan or a policy) is fixed (Fig. 10.1). Starting with a simple "baseline" planning model, how should the subject agent adapt its model to the beliefs and preferences of others in order to improve its planning? Algorithms for this problem setting enjoy many applications. They could help build new and smarter AI embedded in real-time strategy games that repeatedly interacts with the existing AI (whose programmed behavior is usually fixed) learning its strategy and using it in its own planning. The methods also find application in building a smart robotic soccer player that joins an *ad hoc* team of other soccer players with differently programmed play (Albrecht and Ramamoorthy, 2013b; MacAlpine *et al.*, 2014; Stone *et al.*, 2010). On repeatedly interacting with a teammate, subject robot adapts its play to the strengths of the teammate.

We focus on individual planning in multiagent settings as formalized by the graphical interactive dynamic influence diagrams (I-DIDs) (Doshi *et al.*, 2009; Zeng and Doshi, 2012), which are extended from single-agent DIDs (Howard and Matheson, 1984). I-DIDs has been widely accepted as a general framework for sequential decision making in uncertain multiagent settings. Emerging applications in automated vehicles that communicate (Luo *et al.*, 2011), integration with the belief–desire–intention framework (Chen *et al.*, 2013c), and toward *ad hoc* teamwork (Chandrasekaran *et al.*, 2014) motivate advances for I-DIDs.

Expectedly, solving planning problem in multiagent settings tends to be computationally complex due to both the curses of dimensionality and history (Pineau *et al.*, 2006b). In particular, besides the traditional physical states, the candidate behavioral models of other agents, which are infinitely many, are also included in agent's state space. These models could be I-DIDs in lower levels thereby leading to a nested modeling: the other agents also model the subject agent in turn. Along with other agents performing actions, receiving observations, and updating their beliefs, I-DIDs also need to maintain the evolution of the considered models regarding other agents over time. The number of these models grows exponentially, subject to received observations and performed actions, over time also expands the dimensionality of the state space. This is further exacerbated by the nested levels. The explicitly modeling the other agents is computationally expensive, but it is the critical feature that allows I-DIDs different from other frameworks and suitable for both cooperative and competitive multiagent settings. Previous efforts focus on identifying behaviorally equivalent models thereby leading to a suite of techniques for lossless and lossy compressions of the model space in I-DIDs (Zeng *et al.*, 2012). As a large space of distinct behaviors exists, these algorithms must still maintain a significantly large model space and are unable to plan for large horizons.

As I-DIDs are probabilistic graphical models, they compactly represent the problem of how an agent should act in an uncertain environment shared with others of unknown types and provide a way to embed and exploit the inherent structure, including the knowledge about the physical world and the other agents, often presented in real-world decision-making scenarios. More importantly, the graphical representation allows conducting inference based observations online. In this context, the approach presented in this chapter makes the following contributions:

(1) We present an approach that ascribes an arbitrary size-limited set of models to other agents and *adapts* this set using trajectories from online interactions. The approach leads to a novel iterative algorithm, OPIAM, for online planning in settings shared with other agents as delineated above. The approach is also useful in other algorithms that consider models such as point-based methods for solving interactive POMDPs (Doshi and Perez, 2008), memory-bounded dynamic programming for decentralized POMDPs (Seuken and Zilberstein, 2007), and online planning for *ad hoc* teamwork (Wu *et al.*, 2011a).

(2) In a first for online planning in multiagent settings, OPIAM exhibits a provable probabilistic guarantee that the approximation error is bounded. The probabilistic error bound improves as more trajectories are obtained.

(3) On two problem domains with up to five agents, we empirically demonstrate that by considering a limited but adaptive model space, the online planning is faster compared to the previous best compression of model spaces (Zeng *et al.*, 2011b), and simultaneously obtains high average rewards. This makes OPIAM outperform current I-DID algorithms, although it has the benefit of online interactions. We observe in our experiments that the true model or its observationally equivalent counterpart almost always gets included in the limited model space as the interactions progress and the space is updated. This provides an explanation for the good quality behavior despite considering a small model space.

10.2 Individual Decision Making Frameworks

This section will formally introduce the graphical multiagent decision making framework named as I-DID which is regarded as the graphical counterpart of interactive partially observable Markov decision processes. This section will be presented in a two-agent setting: agent

i and j, but all discussion can be extended to many-agent setting straightforwardly.

10.2.1 *Interactive Dynamic Influence Diagrams*

Different from the enumerative representation of decision making problems e.g., POMDPs and interactive POMDPs (I-POMDPs), graphical models, such as influence diagrams (IDs) (Howard and Matheson, 1984; Shachter, 1986) provides a quantitative specification language that decomposes the problem into random variables and dependencies in between. In such formalizations, structures commonly exist in real-world applications can be exploited as conditional independencies between variables to facilitate more efficient reasoning (Boutilier and Poole, 1996). In the rest of this section, we will present more details of this decision-making frameworks with the help of graphical representations.

For a single agent i situated in an uncertain environment, IDs are well suited to describe its one-shot decision-making problem. As a type of probabilistic graphical models, IDs employ chance nodes to model the uncertain aspects of the problem as random variables, such as S modeling the uncertainty regarding the physical environment and O_i modeling agent i's noisy observations (Fig. 10.2(a)). Besides, IDs use a decision node A_i and a utility node R_i to model agent i's actions and associated reward function, respectively. After evaluating

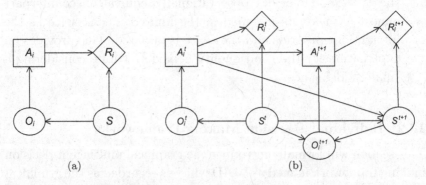

Fig. 10.2.   (a) An ID for single-agent one-shot decision making. (b) A two-slice DID for single-agent sequential decision making.

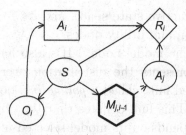

Fig. 10.3. A generic level $l > 0$ I-ID for agent i situated with one other agent j. The highlighted hexagon is the model node, $M_{j,l-1}$, and the dashed arrow is the policy link to other's actions.

the ID for each possible setting of the decision node and state variables, action(s) with the greatest expected utility is are selected as its optimal choice. When agent i sequentially takes actions along time, IDs are extended into dynamic IDs (DIDs). DIDs are regarded as the graphical counterpart of traditional POMDPs, where the transition function is represented by the link between S^t and S^{t+1} and the temporal constrain over actions is represented by the one between A_i^t and A_i^{t+1} (Fig. 10.2(b)).

In order to address decision-making problems in multiagent settings, IDs are extended to interactive IDs (I-IDs) (Doshi *et al.*, 2009). Besides nodes already used in IDs, there is a new type of node, called the *model node* (the hexagonal node in Fig. 10.3), introduced into the I-ID formalism. The model node, as a part of agent i's belief, contains alternative computational models ascribe to the agent j, as agent j's action influences the shared environment, agent i's observation and utility. To be consistent with the I-POMDP framework (Gmytrasiewicz and Doshi, 2005), agent j's level is set one less than that of the subject agent i. That is, given agent i's *strategy level* is l, the agent j is at level $l - 1$. It emphasizes the nested modeling between agents i and j, and it ends with a basis level 0 model which can be an ID or a flat probability distribution over actions (aka, subintentional models). The chance node S and the model node $\mathcal{M}_{j,l-1}$, which contains a set of possible models under consideration, form the *interactive state space* of agent i. Within $\mathcal{M}_{j,l-1}$, each model $m_{j,l-1} = \langle b_{j,l-1}, \hat{\theta}_j \rangle$, has agent j's level $l-1$ belief,

$b_{j,l-1}$, over its interactive state space, and its *frame*, $\hat{\theta}_j$, comprising decision, observation and utility nodes.

In addition to the model node, I-IDs also have an extra chance node, A_j, that represents the distribution over the other agent's actions and a dashed link, called a *policy link*, between model nodes and chance nodes. This link denotes that the object agent's action, indicated in A_j, depends on its model selected in the model node.

The newly introduced model node and the dashed policy link that connects model and chance node can be implemented by ordinary ID nodes and links. As shown in Fig. 10.4, an I-ID (a) is transformed into a flat ID (b). First, the decision of each level $l-1$ I-ID or level 0 ID is mapped into a chance node. Specifically, if $\mathsf{OPT}(m^1_{j,l-1})$ is the set of optimal actions obtained by solving the lower level I-ID (or ID) denoted by $m^1_{j,l-1}$, then the corresponding distribution over the mapped chance node, A^1_j, is: $\Pr(a_j \in A^1_j) = \frac{1}{|\mathsf{OPT}(m^1_{j,l-1})|}$ if $a_j \in \mathsf{OPT}(m^1_{j,l-1})$, 0 otherwise.

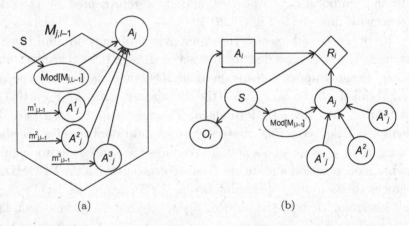

(a)   (b)

Fig. 10.4. (a) The model node and policy link in Fig. 10.3 may be represented using chance nodes and dependencies between the nodes. The decision nodes of the lower-level I-IDs or IDs ($m^1_{j,l-1}, m^2_{j,l-1}, m^3_{j,l-1}$ where the superscript numbers serve to distinguish the models) are mapped to the corresponding chance nodes (A^1_j, A^2_j, A^3_j) respectively, which is indicated by the dotted arrows. Depending on the value of node, $\mathrm{Mod}[M_j]$, distribution of each of the action chance nodes is assigned to node A_j with some probability. (b) The I-ID of Fig. 10.3 transforms into the flat ID with the model node and policy link replaced as in (a).

For each model, say $m^1_{j,l-1}$, there is a corresponding chance node A^1_j. There are as many such action nodes as the number of models in the support of agent i's beliefs, $|\mathcal{M}_{j,l-1}|$. Besides, there is a link from the chance node labeled $\text{Mod}[M_j]$ to the chance node, A_j. The distribution over $\text{Mod}[M_j]$ is i's belief over considered j's models given the state, as a part of i's belief regarding the entire interactive state space. The conditional probability table (CPT) of the chance node, A_j, is a *multiplexer* that selects a distribution of actions nodes (e.g., one out of A^1_j, A^2_j, A^3_j) depending on the value of the selector, $\text{Mod}[M_j]$. In other words, when the belief over $\text{Mod}[M_j]$ is deterministic, e.g., $m^1_{j,l-1}$, the chance node A_j assumes the distribution of the node A^1_j purely; or A_j will take a weighted mixture distribution of all A^*_j based on the distribution over models in $\text{Mod}[M_j]$. Beyond the two-agent setting, for each agent in addition to agent j, we add a model node and a chance node representing the distribution over that agent's action linked together using a policy link.

Example 10.1. The multiagent tiger problem domain involves two closed doors one of which hides a tiger and the other hides a pot of gold (physical states denoted by $S = \{\text{TL}, \text{TR}\}$), and two agents, i and j, which face the closed doors. Each agent may open either the left door (action denoted by OL) or open the right door (OR), or listen (L): $A_i = A_j = \{\text{OL}, \text{OR}, \text{L}\}$.

Let us assume agent i considers agent j. On listening, agents may hear the tiger growling either from behind the left door (observation denoted by GL) or from behind the right door (GR). Additionally, agent i hears creaks emanating from the direction of the door that was possibly opened by agent j. This includes creak from the left (CL), creak from the right (CR), or silence (S) if no door was opened. Formally, $\Omega_j = \{\text{GL}, \text{GR}\}$ and $\Omega_i = \{\text{GLCL}, \text{GLCR}, \text{GLS}, \text{GRCL}, \text{GRCR}, \text{GRS}\}$. If any door is opened by an agent, the tiger persists in its original location with a probability of 95%. All observations are assumed to be noisy. Agent i hears growls with a reliability of 65% and creaks with a reliability of 95%, while agent j hears growls with a reliability of 95%. In other words, agent

i hears agent j's actions more reliable than the tiger's location, and therefore i can use j's actions as an indication of the location of the tiger. An agent gets rewarded for opening the door that hides the gold but gets penalized for opening the door hiding the tiger.

Agent j's models at level 0 can be different, for example, its belief about the tiger's location. Specifically, agent j's models are traditional POMDPs for the single-agent tiger problem, and later included in agent i's interactive state space, $IS_{i,1}$, denoted as $M_{j,0}$. The transition, observation, and reward function can be easily specified based on the previous description. After solving models in $M_{j,0}$, and combining with the agent i's belief regarding j's models, we get the probability distribution of j's actions. Consequently, agent i can reason out its optimal action from its own perspective.

For the one-shot multiagent tiger problem, agent i's decision-making model at level 1 can be represent by the I-ID in Fig. 10.5. Different from the enumerative representation I-POMDP$_{j,0}$

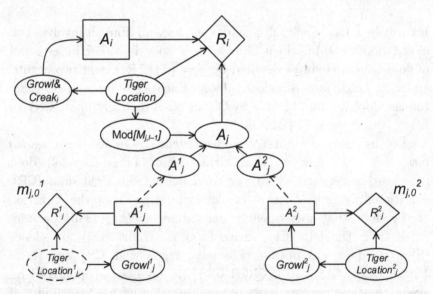

Fig. 10.5. Level 1 I-ID of i for the multiagent tiger problem. Solutions of two-level 0 models (IDs) of j map to the chance nodes, $A_j^{t,1}$ and $A_j^{t,2}$, respectively (illustrated using dotted arrows), transforming the I-ID into a flat ID. The two models differ in the distribution over the chance node, TigerLocation^t.

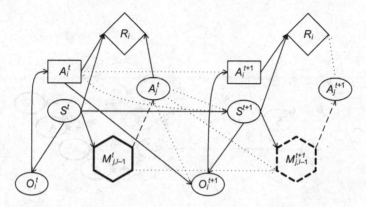

Fig. 10.6.   A generic two time-slice level l I-DID for agent i. The dotted arrow is the model update link that denotes the update of the models of j and of the distribution over the models as both agents act and observe over time.

and I-POMDP$_{i,1}$, each component therein is graphically shown in Fig. 10.5.

When the subject agent i is going to perform a series of actions with the presence of the other agent, I-IDs are further generalized into I-DIDs to address the multiagent sequential decision making problems. In addition to model nodes and policy links, an I-DID includes *model update links*, shown as a dotted arrow in Fig. 10.6, to describe the agent j's model updating over time.

Typically, agents' beliefs are updated subject to performed acts and received observations. Therefore, the update of the model node over time involves two steps. First, given the models at time t, we identify the updated set of models that reside in the model node at time $t+1$. As the agent has the possibility to perform any action and receive any observation, the updated model set at time step $t+1$ can have up to $|\mathcal{M}_{j,l-1}^t||A_j^*||\Omega_j^*|$ models, where $|\mathcal{M}_{j,l-1}^t|$ is the number of models at time step t, $|A_j^*|$ and $|\Omega_j^*|$ are the largest number of actions and observations, respectively, among all the models. Note that, the CPT of chance node, $\text{Mod}[M_{j,l-1}^{t+1}]$, encodes an indicator function, $\tau(b_{j,l-1}^t, a_j^t, o_j^{t+1}, b_{j,l-1}^{t+1})$. The function value is 1 if the belief $b_{j,l-1}^t$ in a model $m_{j,l-1}^t$ using the action a_j^t and observation o_j^{t+1} updates to $b_{j,l-1}^{t+1}$ in a model $m_{j,l-1}^{t+1}$; otherwise it is 0.

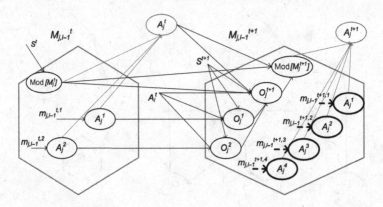

Fig. 10.7. The model update link may be represented using chance nodes and dependency links between them. Consider two models in the model node at time t. These grow *exponentially* in number to more models in the model node at $t+1$ as shown in bold (superscript numbers distinguish the different models). Models at $t+1$ reflect the updated beliefs of j, and their solutions provide the probability distributions for the corresponding action nodes.

Second, the new distribution over the updated models, given the original distribution and the probability of the agent performing the action and receiving the observation that led to the updated model, is computed. The dotted model update link in the I-DID may be implemented using standard dependency links and chance nodes, as shown in Fig. 10.7 thereby the I-DID can be transformed into a flat DID. We show the two time-slice flat DID with the model nodes and the model update link replaced by the chance nodes and the relationships between them, in Fig. 10.9. Chance nodes and dependency links not in bold are standard, usually found in single-agent DIDs.

Example 10.2. We illustrate the two time-slice I-DID for the multiagent tiger problem in Fig. 10.8. The model-update link not only updates the number of j's candidate models due to j's action and observations of growl and creak, it also updates the probability distribution over these models. In Fig. 10.9, we illustrate the update of a single model of j contained in the model node at time t over time.

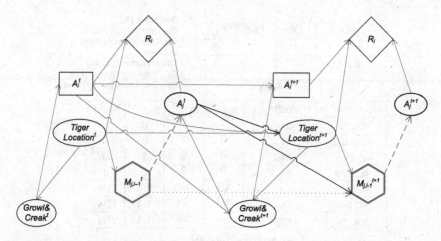

Fig. 10.8.   Two time-slice level l I-DID of i for the multiagent tiger problem. Highlighted model nodes in hexagon-shaped contain the different models of j.

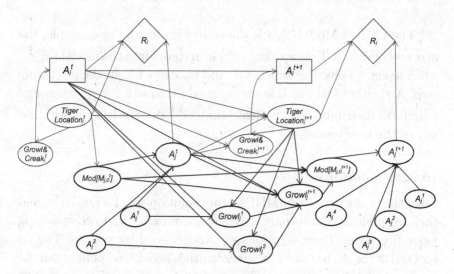

Fig. 10.9.   A flat DID obtained by replacing the model nodes and model-update link in the I-DID of Fig. 10.6 with the chance nodes and the relationships (in bold) as shown in Fig. 10.7. The lower-level models are solved to obtain the distributions for the action chance nodes.

Mod[$M_{j,0}^t$]	OL	OR	L
$m_{j,0}^{t,1}$	1	0	0
$m_{j,0}^{t,2}$	0	0	1

CPT of A_j^t

Decisions (A_j)
OL: Open the left door
OR: Open the right door
L: Listen
Observations (Growl$_j$)
GL: Growl from the left door
GR: Growl from the right door

<A_j^t, Growl$_j^{t+1}$>	Mod[$M_{j,0}^t$]	$m_{j,0}^{t+1,1}$	$m_{j,0}^{t+1,2}$	$m_{j,0}^{t+1,3}$	$m_{j,0}^{t+1,4}$
<OL, GL>	$m_{j,0}^{t,1}$	1	0	0	0
<OL, GR>	$m_{j,0}^{t,2}$	0	1	0	0
<L, GL>	$m_{j,0}^{t,1}$	0	0	1	0
<L, GR>	$m_{j,0}^{t,2}$	0	0	0	1
<OR, *>	*	1/4	1/4	1/4	1/4
<L, *>	$m_{j,0}^{t,1}$	1/4	1/4	1/4	1/4
<OL, *>	$m_{j,0}^{t,2}$	1/4	1/4	1/4	1/4

CPT of Mod[$M_{j,0}^{t+1}$]

Fig. 10.10. The CPTs of the chance nodes A_j^t and Mod[$M_{j,0}^{t+1}$].

The CPT of Mod[$M_{j,0}^{t+1}$] is shown in Fig. 10.10. For example, the first row of the CPT shows that $m_{j,0}^{t,1}$ is updated into the model $m_{j,0}^{t+1,1}$ when agent j takes the action OL and observes GL in the next time step. As neither OR nor L is the optimal decision for $m_{j,0}^{t,1}$, we assign a uniform distribution to indicate that $m_{j,0}^{t,1}$ does not transform into any of the new models for these actions.

10.2.2 *Solutions*

Like the traditional POMDPs, the solution of I-POMDPs and their graphical counterpart I-DIDs maximizes the corresponding value functions. Their solutions are also defined by policies. For an I-POMDP or I-DID with a given initial belief, its policy can be represented by a tree, named as policy tree. Essentially, a policy tree is a regular tree with meaningful labels on nodes and edges where nodes are labeled by optimal actions and edges are labeled by all possible observations. Due to the nested modeling of the other agent, in order to solve the I-POMDP or the I-DID for agent i at level l, all considered possible models of agent j at level $l-1$ need to be solved first. Such recursion ends with the base level 0 model, which can be a traditional POMDP or DID of horizon T. The solutions of level $l-1$ models, e.g., policy trees, are entered in the corresponding action nodes found in the model node of the level l I-DID at the corresponding time step (lines 3–5).

After the model expansion, the solution method uses the standard look-ahead technique, projecting the agent's action and observation sequences forward from the current belief state, and finding the possible beliefs that i could have in the next time step (Russell and Norvig, 2010). Because agent i has a belief over j's models as well, the look-ahead includes finding out the possible models that j could have in the future. Consequently, each of j's level 0 models represented using a standard DID must be solved in the first time step up to horizon T to obtain its optimal policy. These actions are combined with the set of possible observations that j could make in that model, resulting in an updated set of candidate models (that include the updated beliefs) that could describe the behavior of j. $SE(b_j^t, a_j, o_j)$ is an abbreviation for the belief update (lines 8–13). Beliefs over this updated set of candidate models are calculated using the standard inference methods through the dependency links between the model nodes shown in Fig. 10.7 (lines 15–18). Agent i's I-DID is expanded across all time steps in this manner. Because I-DIDs are transformed into flat DIDs, we point out that the Algorithm 10.1 may be realized with the help of standard implementations of DIDs such as HUGIN EXPERT (Andersen and Jensen, 1989). The solution is a *policy tree* that prescribes the optimal action(s) to perform for agent i initially given its belief, and the actions thereafter conditional on its observations up to time T.

Example 10.3. We consider a particular situation of the multiagent tiger problem. Agent i believes agent j has similar reward function and j has superior sensing ability: agent j can correctly figure out tiger's location given one observation with probability 0.95, while agent i can only success with probability 0.65. But agent i can reliably sense j's movement. When agent i does not have any prior information regarding tiger's location, it will be wise to just

---

Algorithm 10.1: Algorithm for exactly solving a level $l \geq 1$ I-DID or level 0 DID expanded over T time steps in a two-agent setting.

Exact Solution (level $l \geq 1$ I-DID or level 0 DID, horizon T)

Expansion Phase
1.   **For** t **from** 0 **to** $T - 1$ **do**
2.   **If** $l \geq 1$ **then**
         $\underline{\text{Minimize } M_{j,l-1}^t}$
3.   **For each** m_j^t **in** $\mathcal{M}_{j,l-1}^t$ **do**
4.   Recursively call algorithm with the $l - 1$ I-DID
              that represents m_j^t and the horizon, $T - t$
5.   Map the solution, $\mathsf{OPT}(m_j^t)$, to the
              corresponding chance node A_j
6. $\mathcal{M}_{j,l-1}^t \leftarrow \mathbf{ModelSpaceCompression}(\mathcal{M}_{j,l-1}^t)$
7.   **If** $t < T - 1$ **then**
         $\underline{\text{Populate } M_{j,l-1}^{t+1}}$
8.   **For each** m_j^t **in** $\mathcal{M}_{j,l-1}^t$ **do**
9.   **For each** a_j **in** $\mathsf{OPT}(m_j^t)$ **do**
10.   **For each** o_j **in** O_j (part of m_j^t) **do**
11.   Update j's belief, $b_j^{t+1} \leftarrow SE(b_j^t, a_j, o_j)$
12. $m_j^{t+1} \leftarrow$ New I-DID (or DID) with b_j^{t+1} as initial belief
13. $\mathcal{M}_{j,l-1}^{t+1} \overset{\cup}{\leftarrow} \{m_j^{t+1}\}$
14.   Add the model node, $M_{j,l-1}^{t+1}$, and the model update link between
            $M_{j,l-1}^t$ and $M_{j,l-1}^{t+1}$
15.   Add the chance, decision, and utility nodes for $t + 1$ time slice and the
        dependency links between them
16.   Establish the CPTs for each chance node and utility node

Solution Phase
17.   **If** $l \geq 1$ **then**
18.   Represent the model nodes, policy links and the model-update links
          as in Fig. 10.7 to obtain the DID
19.   Apply the standard look-ahead and backup method to solve the
        expanded DID (other solution approaches may also be used)

---

follow agent j's action. Agent i considers two models for j, which differ in j's level 0 initial beliefs toward the physical world. In one model, j assigns probability 0.9 that tiger behinds the left door $(\mathrm{Pr}(S_{j,0}^1 = \mathsf{TL}) = 0.9)$, while the other one assigns 0.1 to that location $(\mathrm{Pr}(S_{j,0}^2 = \mathsf{TL}) = 0.1)$. Figure 10.11(a) shows the policy tree of

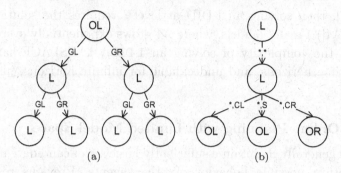

Fig. 10.11. (a) Agent j's policy tree. (b) Agent i's policy tree. We use "*" to denote any one of the observations.

$m_{j,0}^1$, and Figure 10.11(b) shows the one obtained by solving $m_{i,1}$, which is first expanded by $\pi_{m_{j,0}^1}$ and $\pi_{m_{j,0}^2}$ following the procedure in Algorithm 10.1. It can be spotted from $\pi_{m_{i,1}}$ that agent i is not sensitive with its observation about tiger's location, but only act based on the information about agent j's action.

Beyond a framework for single-agent sequential decision-making problems, I-DIDs model the other unknown type agent, j, by including all its possible decision models into the subject agent i's model spaces. Therefore, the scalability of their solutions suffer from not only great number of possible models of agent j, $\mathcal{M}_{j,l-1}^0$. Moreover, the number of such models grows exponentially due to the curse of history (i.e., $|\mathcal{M}_{j,l-1}^0|(|A_j||\Omega_j|)^t$ models to be considered at time step t) and the nested modeling (i.e., the solution of each considered level $l-1$ model requires solving the lower level $l-2$ models, and so on recursively up to level 0). Recall the bottom-up nesting procedure, the 0th level intentional models are DIDs, and after solving these models, their provide policies for agents modeled at that level as subintentional models to a model at level 1. Given these policies, the level 1 model can be solved as flat DIDs again, and further provide policies to models at higher levels. Assume that the number of intentional models considered at each level is bound by a number, $|\Theta|$. Solving an I-DID$_{i,l}$ is then equivalent to solving $O(|\Theta|^l)$ DIDs. For a problem includes N other agents, if the number of models considered at each level for an agent is bound by $|\mathcal{M}|$, then solving an I-DID at level l requires the solutions of $\mathcal{O}((N|\mathcal{M}|)^l)$ many models, where $\mathcal{M}$ grows exponentially over time. Hence, the complexity of solving an I-DID$_{i,l}$ is PSPACE-hard for finite time horizons, and undecidable for infinite horizons, just like for DIDs.

10.3   Online Planning with Limited Model Space

I-DIDs generally maintain a sufficiently large set of candidate models to capture possible behavior of other agents. Previous research focuses on reducing the expanding space of models at every time step (Zeng and Doshi, 2012). Nevertheless, the methods are challenged by the *exponential growth* in the number of models over time. Moreover, the existing *de facto* method for updating distributions over the models is to use a Bayesian update that does not adapt the model space-prune or replace candidate models.

Using the insight that *observations received by agent i can reveal the actions performed by agent j which depend on its model*, we propose a new algorithm that exploits agents' interactions to improve I-DID solutions by focusing on identifying agent j's true behavioral model. We outline the algorithm and discuss the steps.

10.3.1   *Algorithm Outline*

Methods that model others operate in the context of *two* sets of models: a large universal set of models and the model set considered by the method. Given agent i's belief distribution over a set of j's models, $b_{i,l}(\mathcal{M}_{j,l-1})$, previous algorithms (Zeng and Doshi, 2012) solve i's I-DID with the *complete* set of j's models. Let $\mathcal{M}_{j,l-1}^0$ denote the set of j's models included in the model node of the I-DID initially. Then, in these approaches $\mathcal{M}_{j,l-1}^0 = \mathcal{M}_{j,l-1}$. Subsequently, agent i interacts with j using the policy obtained by solving the I-DID and empirically updates $b_{i,l}(\mathcal{M}_{j,l-1})$ accordingly. Doshi and Gmytrasiewicz (2006) show that $b_{i,l}(\mathcal{M}_{j,l-1})$ converges to a probability distribution, denoted by $b_{i,l}^*(\mathcal{M}_{j,l-1})$, after a sufficiently large number of interactions if the initial belief satisfies the absolute continuity condition.

In contrast, our new algorithm, OPIAM (**Online plan, interact and adapt models**), solves an initial "baseline" I-DID using a *partial* set of j's models: $\mathcal{M}^0_{j,l-1} \subset \mathcal{M}_{j,l-1}$. Agent i then uses the solved policy to interact with j for some time and updates $b_{i,l}(\mathcal{M}^0_{j,l-1})$ using the trajectories of its actions and observations obtained from the interactions. Subsequently, OPIAM modifies the I-DID by partially replacing $\mathcal{M}^0_{j,l-1}$ with models from the unexplored set $(\mathcal{M}_{j,l-1}/\mathcal{M}^0_{j,l-1})$.

We present OPIAM in Fig. 10.12. Given a large set of models of agent j, $\mathcal{M}_{j,l-1}$, we randomly select a *small* set of candidate models, $\mathcal{M}^0_{j,l-1}$, and initialize the I-DID, $m_{i,l}$, with a uniform distribution over the models (lines 1–2). Solving $m_{i,l}$ provides an initial policy that agent i executes to play with agent j online (line 6). Formally, we denote a T-horizon policy as $\pi^T_{m_{i,l}}$ that is represented using a tree and contains a set of policy paths from the root node to the leaf.

Definition 10.1 (Policy path). A policy path, $h_i^{T,(k)}$, is an action-observation sequence of T steps: $h_i^{T,(k)} = \{a_i^t, o_i^{t+1}\}_{t=0}^{T-1}$, where o_i^T is null.

After an interaction, agent i weights each of j's candidate model, $\Pr(m_{j,l-1}|h_i^{T,(k)})$, given i's observations and executed actions up to T steps (line 7). This computation uses Bayes rule in a straightforward way:

$$\Pr(m_{j,l-1}|h_i^{T,(k)}) \propto \sum_{s^0} \Pr(m_{j,l-1}|s^0)\Pr(h_i^{T,(k)}|m_{j,l-1}, s^0)$$

where $\Pr(m_{j,l-1}|s^0)$ is the prior weight of j's model given the state and $\Pr(h_i^{T,(k)}|m_{j,l-1}, s^0)$ the likelihood of i's policy path. We compute this by inserting $h_i^{T,(k)}$ as evidence into the corresponding decision and chance nodes, and $m_{j,l-1}$ in the model node of the level l I-DID, $m_{i,l}$.

OPIAM(level $l \geq 1$ I-DID of agent i, $m_{i,l}$; candidate
    models of agent j at level $l-1$ or level 0, $\mathcal{M}_{j,l-1}$; horizon T; ρ)

1. Weight $\mathcal{M}_{j,l-1}$ equally and set the counter, $\tau = 1$
2. Select a subset of j's candidate models, $\mathcal{M}_{j,l-1}^0 \subset \mathcal{M}_{j,l-1}$
3. **Do**
4.   Solve agent i's model, $m_{i,l}$ with $\mathcal{M}_{j,l-1}^0$, using
      I-DID Exact in Fig. 10.11, and output i's policy $\pi_{m_{i,l}}$
      <u>Interaction Phase</u>
5.   **For** $k = 1$ to N interactions of length T **do**
6.   Agent i plays with j according to the policy $\pi_{m_{i,l}}$
7.   Compute the posterior weights of selected j's models
        $\Pr\left(m_{j,l-1}|h_i^{T,(k)}\right)$
8.   Compute j's most probable path ξ_j^T and
        update its occurrence ω_{ξ_j}
9.   **For each** $m_{j,l-1}$ in $\mathcal{M}_{j,l-1}^0$ **do**
10.   Average the posterior weight: $\sum_{k=1}^{N} \Pr\left(m_{j,l-1}|h_i^{T,(k)}\right)/N$
11.   Update beliefs about the selected models $b_{i,l}^\tau(\mathcal{M}_{j,l-1}^0)$ using Eq. 10.1
12.   Calculate $\Delta = \|b_{i,l}^\tau(\mathcal{M}_{j,l-1}^0) - b_{i,l}^{\tau-1}(\mathcal{M}_{j,l-1}^0)\|_2$
13.   **If** $\Delta \leq \rho$ **then**
14.   Set $\tau = 0$
15.   **else**
16.   Set $\tau = \tau + 1$
      <u>Model Adaptation Phase</u>
17.   Aggregate most probable paths ξ_j^T into $\mathcal{H}_j^T$
18.   Select the most probable model, $\dot{m}_{j,l-1} \in \mathcal{M}_{j,l-1}/\mathcal{M}_{j,l-1}^0$
19.   Select model, $m_{j,l-1} \in \mathcal{M}_{j,l-1}^0$, with the least
      average weight
20.   Replace $m_{j,l-1}$ with $\dot{m}_{j,l-1}$ in $\mathcal{M}_{j,l-1}^0$
21. **While** $\tau > 0$

Fig. 10.12. OPIAM in combination with a level l I-DID expanded over T steps allows online planning with a limited model space.

After N interactions, the average weight of each model is obtained, $\sum_{k=1}^{N} \Pr(m_{j,l-1}|h_i^{T,(k)})/N$ (line 10). As not all of j's models are included in the current I-DID, **OPIAM** next updates the belief over the partial set of models in the larger model space $\mathcal{M}_{j,l-1}$, denoted by $b_{i,l}^\tau(\mathcal{M}_{j,l-1}^0)$. This is done by redistributing the probability mass in i's belief for the selected models in $\mathcal{M}_{j,l-1}^0$ over all j's models in $\mathcal{M}_{j,l-1}^0$ proportionally to their average weights obtained in the previous step. This is shown in Eq. (10.1):

$$b_{i,l}^{\tau}(m_{j,l-1}) = \frac{1}{N} \sum_{k=1}^{N} \Pr(m_{j,l-1}|h_i^{T,(k)})$$

$$\times \sum_{m_{j,l-1} \in \mathcal{M}_{j,l-1}^0} b_{i,l}^{\tau-1}(m_{j,l-1}) \qquad (10.1)$$

where the second factor that sums $b_{i,l}^{\tau-1}(m_{j,l-1})$ over all models in $\mathcal{M}_{j,l-1}^0$ is the total probability mass over models included in the I-DID, and $\Pr(m_{j,l-1}|h_i^{T,(k)})$ is the updated weight of model $m_{j,l-1}$ given i's policy path, $h_i^{T,(k)}$.

Subsequently, OPIAM prunes the model in $\mathcal{M}_{j,l-1}^0$ with the lowest weight and replaces it with one from the remaining models (lines 19–20). The initial belief in the I-DID for the next round of interactions is a normalized distribution over the updated $\mathcal{M}_{j,l-1}^0$. We terminate model adaptation when the change in the belief distribution is less than a small value, ρ, during two consecutive iterations (lines 12–14).

In principle, the online algorithm allows agent i to explore j's possible models with the overall goal of identifying j's true behavior. Next, we present a method for selecting a model in $\mathcal{M}_{j,l-1}$ as replacement in the *model adaptation* phase (lines 17–18).

10.3.2 *Most Probable Model Selection*

A key observation is that while the true model of agent j cannot be directly observed, j's behavior induces i's observations, which in turn provide inferential information about j's models. Hence, we focus on the challenge of selecting the most probable model of j based on i's action–observation history for adapting model space, $\mathcal{M}_{j,l-1}^0$.

We point out that every interaction involves only one particular policy path of j. After every interaction, agent i infers the most probable policy path for agent j given i's actions and observations. Formally, we define the most probable path below.

Definition 10.2 (Most probable path). Given agent i's model, $m_{i,l}$, and i's sequence of actions and observations, $h_i^{T,(k)}$, define the most probable path for agent j as:

$$\xi_j^{T,(k)} = \underset{h_j^T \in H_j^T}{\arg\max} \; \Pr(h_j^T | h_i^{T,(k)})$$

$$= \underset{h_j^T \in H_j^T}{\arg\max} \; \prod_{t=1}^{T-1} \Pr(a_j^t | h_i^{T,(k)}, h_j^{t-1}, o_j^t) \, \Pr(o_j^t | h_i^{T,(k)}, h_j^{t-2})$$

$$\times \Pr(a_j^0 | h_i^{T,(k)})$$

The computation factorizes the joint probability $\Pr(h_j^T, h_i^{T,(k)})$ given the graphical structure of the I-DID, and is carried out through a usual evidence propagation in the level l I-DID. The value of each term above is obtained from the distributions in nodes in the I-DID. Here, $H_j^T = A_j \times \prod_{t=1}^{T-1}(\Omega_j \times A_j)$ are all possible policy paths of agent j of T steps as in Definition 10.1.

Because agent j's set of possible paths is large, we may select the most probable path approximately by sampling the most probable action and observation at every time step. Specifically, we compute the most probable action by maximizing $\Pr(a_j^0 | h_i^{T,(k)})$ at time $t = 0$, and subsequently sample the most probable observation and actions over time.

Let $\mathcal{H}_j^T$ be the set of the most probable paths after N interactions, $\mathcal{H}_j^T = \bigcup_{k=1}^N \xi_j^{T,(k)}$. Subsequently, $\mathcal{H}_j^T$ will compose the entire policy tree that j is using, $\pi_{m_j^*}^T$, if agent i interacts with j for a sufficiently long time, i.e., $N \to \infty$.

Consider a candidate model, $m_{j,l-1} \in \mathcal{M}_{j,l-1}$, whose solution is a T-step policy tree, $\pi_{m_j}^T$. Let $\Pr(a_j^t | \pi_{m_j}^T)$ be the proportion among all actions that action, a_j, appears at time step t in all paths of the policy tree. *We seek a measure of how well the candidate model, $m_{j,l-1}$, fits the inferred most probable action-observation histories of j.* Let δ^T denote this measure and we define it as:

$$\delta^T(m_{j,l-1}, \mathcal{H}_j^T) \triangleq \sum_t \sum_{a_j \in A_j} |\Pr(a_j^t | \pi_{m_j}^T) - \Pr(a_j^t | \mathcal{H}_j^T)| \qquad (10.2)$$

where

$$\Pr(a_j^t | \mathcal{H}_j^T) \triangleq \sum_{\xi_j^{T,(k)} \in \mathcal{H}_j^T} \omega_{\xi_j^T} \cdot \Pr(a_j^t | \xi_j^{T,(k)}).$$

Here, $\Pr(a_j^t | \xi_j^{T,(k)})$ is 1 if action a_j appears in the most probable path, $\xi_j^{T,(k)}$, at time step t and weight $\omega_{\xi_j^T}$ is the normalized occurrence of $\xi_j^{T,(k)}$ over N interactions.

Subsequently, we obtain the most probable model for replacement as shown below.

Definition 10.3 (Most probable model). Given the set of agent j's models, $\mathcal{M}_{j,l-1}$, and most probable paths, $\mathcal{H}_j^T$, define the most probable model, $\dot{m}_{j,l-1}$, for the level $l-1$ agent j as:

$$\dot{m}_{j,l-1} = \operatorname*{arg\,min}_{m_{j,l-1} \in \mathcal{M}_{j,l-1}/\mathcal{M}_{j,l-1}^0} \delta^T(m_{j,l-1}, \mathcal{H}_j^T)$$

We elaborate the computation of the most probable model using an example in the multiagent tiger problem (Gmytrasiewicz and Doshi, 2005) below.

Example 10.4. Assume that the initial model set, $\mathcal{M}_{j,l-1}^0$, comprises three models, $\{m_j^1, m_j^2, m_j^3\}$, and another candidate model, m_j^4, is available to be selected for updating $\mathcal{M}_{j,l-1}^0$. Figure 10.13 shows a combination of policy trees ($T = 3$) obtained by solving every model in $\mathcal{M}_{j,l-1}^0$ (a), and the policy tree of m_j^4 (b). After $N = 6$ interactions, $\mathcal{H}_j$ consists of three policy paths: $\xi_1 = \langle \mathrm{OL}, \mathrm{GR}, \mathrm{L}, \mathrm{GR}, \mathrm{OL} \rangle$, $\xi_2 = \langle \mathrm{L}, \mathrm{GL}, \mathrm{L}, \mathrm{GR}, \mathrm{L} \rangle$ and $\xi_3 = \langle \mathrm{L}, \mathrm{GL}, \mathrm{L}, \mathrm{GL}, \mathrm{OR} \rangle$. These are inferred with the (normalized) frequencies, $\omega_{\xi_1} = 1/6$, $\omega_{\xi_2} = 2/3$, and $\omega_{\xi_3} = 1/6$, respectively.

$\Pr(a_j^t | \pi_{m_j^4})$ at the different time steps is given below. The superscripts on actions denote the time step.

$\Pr(\mathrm{L}^1 | \pi_{m_j^4}) = 1$; $\Pr(\mathrm{OL}^1 | \pi_{m_j^4}) = 0$; $\Pr(\mathrm{OR}^1 | \pi_{m_j^4}) = 0$;

$\Pr(\mathrm{L}^2 | \pi_{m_j^4}) = 1$; $\Pr(\mathrm{OL}^2 | \pi_{m_j^4}) = 0$; $\Pr(\mathrm{OR}^2 | \pi_{m_j^4}) = 0$;

$\Pr(\mathrm{L}^3 | \pi_{m_j^4}) = 1/2$; $\Pr(\mathrm{OL}^3 | \pi_{m_j^4}) = 1/4$; $\Pr(\mathrm{OR}^3 | \pi_{m_j^4}) = 1/4$

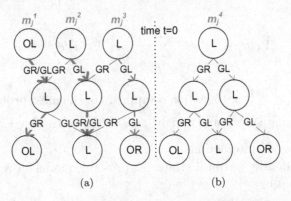

Fig. 10.13. (a) Policy trees of initial models, $\mathcal{M}^0_{j,l-1}$ and (b) the candidate model, m^4_j. Node and edge labels are j's actions and observations respectively. Darker (gray) edges indicate the occurrence of most probable paths over interactions.

Next, we compute $\Pr(a_j|\mathcal{H}_j)$ for each action at each time step.

$$\Pr(L^1|\mathcal{H}_j) = 5/6; \ \Pr(OL^1|\mathcal{H}_j) = 1/6; \ \Pr(OR^1|\mathcal{H}_j) = 0;$$

$$\Pr(L^2|\mathcal{H}_j) = 1; \ \Pr(OL^2|\mathcal{H}_j) = 0; \ \Pr(OR^2|\mathcal{H}_j) = 0;$$

$$\Pr(L^3|\mathcal{H}_j) = 4/6; \ \Pr(OL^3|\mathcal{H}_j) = 1/6; \ \Pr(OR^3|\mathcal{H}_j) = 1/6$$

Finally, the distance, δ, between the two distributions gives a measure of closeness of m^4_j to the observed trajectories:

$$\delta = (|1 - 5/6| + |0 - 1/6|) + (|1 - 1|)$$

$$+ (|1/2 - 2/3| + |1/4 - 1/6| + |1/4 - 1/6|) = 2/3$$

10.4 Savings and PAC Bound

It is well known that the primary complexity of solving I-DIDs is due to the exponentially growing number of j's models over time. At time step t, there could be $|\mathcal{M}^0_{j,l-1}|(|A_j||\Omega_j|)^t$ many models of the other agent j, where $|\mathcal{M}^0_{j,l-1}|$ is the number of models considered initially. Previous approaches consider the entire candidate set of j's models, $\mathcal{M}^0_{j,l-1} = \mathcal{M}_{j,l-1}$, where $\mathcal{M}_{j,l-1}$ is generally large as it seeks to cover as much as feasible possible models of j.

In contrast, OPIAM considers a relatively small set of j's models initially and iteratively explores the rest online. While the adaptation phase eventually solves all models in the initial $\mathcal{M}_{j,l-1}$ (Eq. (10.2)), the primary computational savings arise in the state space during the look ahead expansion: $|\mathcal{M}^0_{j,l-1}|(|A_j||\Omega_j|)^t \ll |\mathcal{M}_{j,l-1}|(|A_j||\Omega_j|)^t$, because $|\mathcal{M}^0_{j,l-1}| \ll |\mathcal{M}_{j,l-1}|$. This makes solving I-DIDs faster — facilitating real-time solution constraints — and scales up the solution in the planning horizon.

Recall that we adapt the model space by including the most probable model. As indicated in Definition 10.3, the model $\dot{m}_{j,l-1}$ is likely to be selected when its predicted actions, $\Pr(a_j|\dot{m}_{j,l-1})$, are similar to those found in the most probable paths, $\Pr(a_j|\mathcal{H}^T_j)$, in the comparison between policy paths. Hence, exploring j's model space online and including the most probable model reduces agent i's loss in value due to misprediction. We take further steps to bound such loss after each iteration of N interactions.

We consider the (typical) condition that agent j's true model $m^*_{j,l-1}$ is part of $\mathcal{M}_{j,l-1}$. Let η be the worst L_1-norm error in the prediction of j's actions at any time step due to the model replacement, $\eta = \max_{t \in T} \sum_{a_j \in A_j} |\Pr(a^t_j|\dot{m}_{j,l-1}) - \Pr(a^t_j|m^*_{j,l-1})|$. The expected value of agent i's optimal policy given by the I-DID is:

$$V^T(m_{i,l}) = \rho(b_{i,l}, a^*_i) + \sum_{o_i} \Pr(o_i|b_{i,l}, a^*_i) V^{T-1}(m'_{i,l})$$

where

$$\rho(b_{i,l}, a^*_i) = \sum_{s, m_{j,l-1}} b_{i,l}(s, m_{j,l-1}) \sum_{a_j} R_i(s, a^*_i, a_j)$$
$$\times \Pr(a_j|m_{j,l-1}).$$

Here, a^*_i is i's optimal action and $m'_{i,l}$ is the updated model of agent i containing the updated belief at the next time step. We denote the expected value and the immediate expected reward of agent i's optimal policy when j's predicted behavior is $\Pr(a_j|\dot{m}_{j,l-1})$, instead of $\Pr(a_j|m^*_{j,l-1})$, as $\dot{V}^T(m_{i,l})$ and $\dot{\rho}(b_{i,l}, a_i)$, respectively. Denote the maximal reward value in the reward function as $R^{\max}_i$.

Proposition 1 with help from Lemmas 1.1 to Lemma 1.4 gives the difference in the value functions due to the misprediction, η. Denote the maximal reward value in the utility function as $R^{\max}_i$.

Lemma 10.1. *For any i's belief, $b_{i,l}$, and action, a_i,*

$$|\rho(b_{i,l}, a_i) - \dot{\rho}(b_{i,l}, a_i)| \leq \eta R_i^{\max}$$

Lemma 10.2. *For any, $b_{i,l}$, a_i, o_i, and a_j, $|Pr(o_i|b_{i,l}, a_i) - \dot{Pr}(o_i|b_{i,l}, a_i)| \leq \eta|\Omega_j|$.*

Lemma 10.3. *For any pair of beliefs, $b'_{i,l}$ and $\dot{b}'_{i,l}$, obtained by updating the same initial belief and the latter obtained by considering j's most probable models for predicting its actions,*

$$Pr(o_i|b_{i,l}, a_i)|b'_{i,l} - \dot{b}'_{i,l}|_1 \leq 2\eta|\Omega_j|$$

Lemma 10.4. *For any pair of updated models of agent i, $m'_{i,l}$, and $\dot{m}'_{i,l}$, where the latter is due to differing predictions of j's actions,*

$$|Pr(o_i|b_{i,l}, a_i)\, V^{T-1}(m'_{i,l}) - \dot{Pr}(o_i|b_{i,l}, a_i)\, V^{T-1}(\dot{m}'_{i,l})|$$
$$\leq 3\eta(T-1)\, R_i^{\max}|\Omega_j|$$

We use Lemmas 10.1 to 10.4 to establish Proposition 1.

Proposition 1. *For a given I-DID with initial belief of agent i, $m_{i,l}$, let $V^T(m_{i,l})$ be the expected reward from solving a T-horizon I-DID optimally, and $\dot{V}^T(\dot{m}_{i,l})$ be the expected reward from solving a T-horizon I-DID by considering the most probable model instead of the true model in the set. Let $\mathcal{E}^T \triangleq |V^T(m_{i,l}) - \dot{V}^T(\dot{m}_{i,l})|$. Then,*

$$\mathcal{E}^T \leq \eta R_i^{\max}((T-1)(1 + 3(T-1)|\Omega_i||\Omega_j|) + 1)$$

Proposition 1 does not as yet bound the error because a non-trivial bound for η is not established. Proposition 2 provides the crucial missing piece probabilistically bounding η using δ^T and an error term, ϵ, that is flexible.

Proposition 2. *$\eta \leq (\delta^T + \epsilon)$ with probability at least $1 - \dfrac{|A_j|T}{e^{2NT(\epsilon/|A_j|T)^2}}$.*

Together, Propositions 1 and 2 establish a PAC bound on the worst error incurred by OPIAM after a single iteration of N interactions between agents i and j. The error accumulates with more iterations until the algorithm terminates due to which the overall bound on the error of OPIAM is loose. For the pathological condition when $m^*_{j,l-1} \notin \mathcal{M}_{j,l-1}$, the probability is not guaranteed to improve with increasing samples unless $b_{i,l}(\mathcal{M}_{j,l-1})$ continues to satisfy the absolute continuity condition (Kalai and Lehrer, 1993).

10.5 Experimental Results

We implemented OPIAM (Fig. 10.12) to improve planning in level 1 I-DIDs. As a baseline, another model replacement technique (replacing line 18 in Fig. 10.12) that *randomly* selects a new model from $\mathcal{M}_{j,l-1}/\mathcal{M}^0_{j,l-1}$, labeled as ROnline is used. Previous best approach for solving I-DIDs compares partial policy trees and belief distributions at the leaf nodes for approximate equivalence and clustering models (Zeng *et al.*, 2012). This method, labeled as ϵ-BE, compacts the complete model space, $\mathcal{M}_{j,l-1}$; it is also allowed to interact and update priors. This approach serves as a high-quality benchmark. A non-adaptive exact method, DMU (Zeng and Doshi, 2012), serves to demonstrate the benefits of adaptation. All experiments are run on a Windows platform with 1.9 GHz i3 processor and 6 GB memory.

We compare their performances on two non-cooperative problem domains: the small multiagent tiger ($|S|=2$, $|A_i|=|A_j|=3$, $|\Omega_i|=6$, $|\Omega_j|=3$, $T=6$) and the larger multiunmanned aerial vehicle (multiUAV) reconnaissance problem ($|S|=25$, $|A_i|=|A_j|=5$, $|\Omega_i|=|\Omega_j|=5$, $T=4$) (Zeng and Doshi, 2012). Models of other agents are IDs that encode the problem and differ in their initial beliefs. Though this space is continuous, BE offers a way to make this space discrete and include the true model. Instead, we evaluated all algorithms for two arbitrary sets of candidate models — $\mathcal{M}_j = 50$ or 100 — from which we select different smaller sets of initial models ($\mathcal{M}^0_j$). This allowed us to experiment with settings where the true models of others are not in $\mathcal{M}_j$. The parameter ρ that guides the termination of the algorithm (Fig. 10.12, line 13) is set as 0.01. Proposition 2 provides a way to obtain N given T, ϵ, and least probability. This resulted in $N \approx 100$ ($T = 6$, $|A_j|=3$, $\epsilon=0.15$ and least probability 0.9), and it varies with T.

We evaluate OPIAM's performance under two conditions: (i) A typical case when the true model, $m^*_{j,l-1} \in \mathcal{M}_{j,l-1}$; and ($ii$) Because $\mathcal{M}_{j,l-1}$ is itself finite in practice, a pathological and bold case where $m^*_{j,l-1} \notin \mathcal{M}_{j,l-1}$. The latter is often utilized to test methods for *ad hoc* cooperation.

For the former case, Fig. 10.14(a) demonstrates the probability mass on the true model with time for different configurations of $\mathcal{M}^0_{j,l-1}$ and $\mathcal{M}_{j,l-1}$ in the tiger problem. Increasing weights indicate that $m^*_{j,l-1}$ enters the smaller model set. More importantly, this model receives much more attention from agent i over time than its initial weight. Here, time is a function of increasing iterations and interaction lengths. Observe that the model weight stabilizes and the difference in distributions approaches zero with the algorithm terminating.

Simultaneously, Fig. 10.14(b) shows the reward obtained over T steps averaged over N. This increases with time indicating that OPIAM is progressively generating higher-valued policies that better predict agent j's actions. ROnline's performance is uneven on both the true model probability and average rewards. Importantly, OPIAM is faster than ϵ-BE whose performance also improves with time due to improved priors, but holding the entire model space slows it down considerably despite its aggressive compression. Each point on the curves indicates a model space adaptation (by OPIAM and ROnline) and prior update (by OPIAM, ROnline and ϵ-BE). Generally fewer points for ϵ-BE suggests that it is slow, despite its compression, in generating a new plan with updated priors compared to others.

For the latter case, Fig. 10.15 (a) shows the differences in model weights over time, as mentioned in line 12 of Fig. 10.12. The difference eventually approaches zero and stays there. The average rewards do not improve as steadily as they do in the previous case, and they reach values that are close to positive but lower than those when the true model is present in the space. However, OPIAM's performance

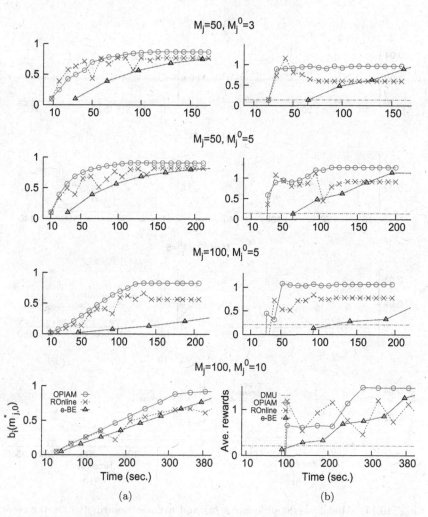

Fig. 10.14. Increasing model weights on the true model (a) and improving average rewards (b) over time for the case: $m^*_{j,l-1} \in \mathcal{M}_{j,l-1}$, in the **multiagent tiger** problem. The horizontal line denotes the average reward from the exact DMU method given a uniform prior over $\mathcal{M}_{j,l-1}$.

remains significantly better than ϵ-BE, ROnline, and DMU. This is insightful revealing that a combination of models, which can be seen as an approximation, may partly compensate for the absence of the true model from the considered space.

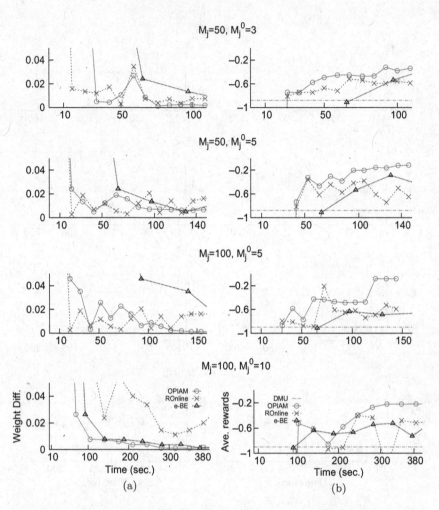

Fig. 10.15. Model weight differences (a) and average rewards (b) for the case: $m_j^* \notin \mathcal{M}_j$, in **tiger**. The weight difference, Δ in line 12 of Fig. 10.12 reaches zero.

In Figs. 10.16 and 10.17, we show the performance of OPIAM on similar metrics in the context of the larger multiUAV problem for the two cases, respectively. While the time duration has increased, OPIAM continues to exhibit average rewards that improve on both ϵ-BE and ROnline. In this larger problem domain, ϵ-BE carried out much less rounds of prior update in the given time period. In some cases, the interaction based on ϵ-BE takes too much time to be shown

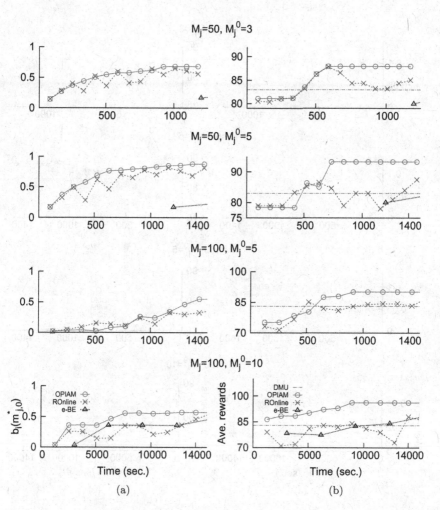

Fig. 10.16. Most probable model weights (a) and average rewards (b) for the case: $m_j^* \in \mathcal{M}_j$, in the larger **multiUAV** problem.

in the plot. Importantly, the true model enters $\mathcal{M}_{j,l-1}^0$ when it's included in the larger set. But, in its absence, the model weight difference approaches zero and stays there. Helped by the online interactions, both OPIAM and ϵ-BE reach rewards that improve on the exact DMU indicated by the flat line.

Finally, we demonstrate the speed up that OPIAM brings in problems with longer planning horizons and when more agents are

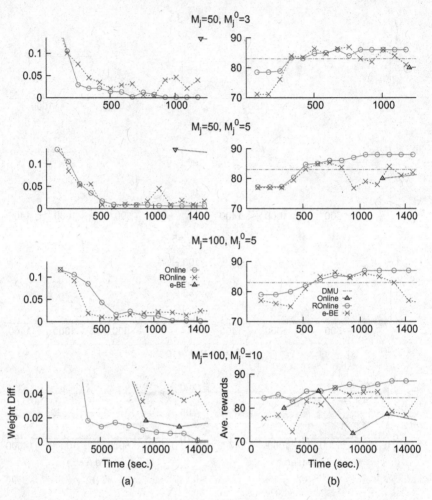

Fig. 10.17. Weight differences over time (a) and average rewards (b) for the case: $m_j^* \notin \mathcal{M}_j$, in **UAV** problem.

sharing the setting. Table 10.1 shows significant speed up compared to ϵ-BE; speed up that increases with problem sizes. ϵ-BE exhibits poor scalability with more agents and the corresponding speed ups improve to multiple orders of magnitude. Such speed ups gradually bring online planning in partially observable multiagent settings to the realistic time requirements imposed on these algorithms.

Table 10.1. OPIAM exhibits significant speed up compared to ϵ-BE with comparable average rewards. These times reflect multiple rounds until convergence.

	Two-agent Tiger (sec)				Two-agent UAV (sec)		
T	ϵ-BE	OPIAM	Speedup	T	ϵ-BE	OPIAM	Speedup
8	65	11.7	5.6	6	2,973	106	28
15	217	67	3.2	8	*	836	n.a.
	Five-agent Tiger (sec)				Four-agent UAV (sec)		
T	ϵ-BE	OPIAM	Speedup	T	ϵ-BE	OPIAM	Speedup
3	13,818	590	23	2	29,638	840	35
6	*	8,394	n.a.	4	*	7,840	n.a.

Note: * indicates the program didn't stop after one-day running.

10.6 Related Work

In settings where others' actions are not directly observable but may be inferred through state changes, Sonu and Doshi (2013) present a bimodal approach in which the subject agent initially uses a single-agent POMDP controller, and switches online to a multiagent controller when it is sufficiently certain of the physical state. With the help of graphically represented interaction structure, the planning problem of agent population has been efficiently addressed (Sonu *et al.*, 2017). Unlike OPIAM, planning is offline and model space is not updated based on data.

While approaches such as OPIAM that target settings shared with other agents with possibly conflicting preferences have been sparse, multiple approaches of online planning for cooperation exist. Wu *et al.* (2011b) use policy equivalence to reduce the size of histories so that agents can continue to coordinate under the condition of limited communication. A second approach, OPAT, by Wu *et al.* (2011a) performs online planning for *ad hoc* teams. It targets a simpler problem setting where the state and joint actions are fully observable. However, Chandrasekaran *et al.* (2014) show that extending OPAT to partial observability did not demonstrate better performance than an

I-DID based solution when the type of the other agent is not known *a' priori*. Harsanyi–Bellman *Ad Hoc* Coordination (HBA) (Albrecht and Ramamoorthy, 2013a) provides a generalized stochastic game framework that maintains a distribution over a set of user-defined teammate types. Reinforcement learning is utilized to learn the agent's optimal actions online. However, similar to OPAT, HBA targets settings where the states and joint actions of others are perfectly observed.

Different from existing online POMDP-based methods (Ross *et al.*, 2008), OPIAM targets multiagent settings and *adapts* the model space thereby repeatedly changing the state space of the problem. OPIAM's focus on a limited set of others' models is reminiscent of point-based value iteration for POMDPs (Pineau *et al.*, 2006b) and interactive POMDPs (Doshi and Perez, 2008). In these techniques, limited belief points are generated from simulated trajectories of the subject agent. However, a key difference is that OPIAM selects models based on probabilistic observations by the subject agent of *others'* trajectories, which help build the most probable paths of others. As we mentioned previously, OPIAM's approach could also be utilized in other algorithms that consider models such as memory-bounded dynamic programming for decentralized POMDPs (Seuken and Zilberstein, 2007).

A predominant factor in the complexity of I-DIDs is due to the exponential growth in the candidate models over time. We may preemptively avoid expanding models that will turn out to be BE to others in the next time step (Doshi and Zeng, 2009; Zeng and Doshi, 2012). By discriminating between model updates, the approach generates a minimal set of models in each non-initial model node.

While exploiting behavioral equivalence (BE) makes the general space of models parsimonious, identifying that two models are BE requires comparing their solutions, which are often policy trees. The trees grow exponentially in size with the horizon of decision making. In a different approach, Zeng *et al.* (2011a) sought to cluster models by comparing the K-most probable paths in the policy trees.

However, computing path probabilities becomes computationally hard as the paths become longer, and bounding the prediction error is not possible. We may further prune the model space by clustering models whose predicted actions at a particular time step are identical (Zeng and Doshi, 2009, 2012). While the clusters may change at each time step, the benefit is that the number of clusters are bounded by the total number of actions of the agent, leading to a small model space. Various I-DID solution techniques that exploit the notion of equivalence were recently compared (Zeng and Doshi, 2012) and their effectiveness demonstrated on several problem domains including in GaTAC. Recently, Conroy *et al.* (2015) starts looking into learning agents behavior from available data, which provide prior knowledge on refining model space in I-DIDs, and a new equivalence principle named as value equivalence (VE) starts attracting research efforts (Conroy *et al.*, 2016a, 2016b).

10.7   Concluding Remarks

A key contribution of **OPIAM** is its capability to work with a small model space and adapt it in order to perform fast online planning. Empirical evaluations demonstrate that average rewards improve with time. In particular, the performance reveals an important insight: when the true model is absent even from the larger model space, combinations of models could approximately fit observed trajectories in a form of equivalence. This is evident from improving rewards in both the problems for this pathological case. Consequently, **OPIAM** represents a significant pragmatic step toward individual online planning in multiagent settings with applications in *ad hoc* cooperation and other methods that maintain models. Our future work involves exploring other ways of adapting the model set that exhibits improved accuracy and demonstrating the benefit of this approach in cooperative settings as well that ascribe models to others. The greater speedups in larger domains also encourage further investigation into representation and solution techniques for I-DIDs.

Acknowledgments

This research is partly supported by NSFC 61502322, 61562033, 61772442. Yingke acknowledges the grant 2017JY0258 from Sichuan Province, China. Yinghui acknowledges the support provided by the grant 20151BAB207021 and 20151BDH80014 from Jiangxi Province, China. This book chapter has been extended from the conference paper: Yingke Chen, Prashant Doshi, Yifeng Zeng, Iterative Online Planning in Multiagent Settings with Limited Model Spaces and PAC Guarantees. In Proceedings of the International Conference on Autonomous Agents and MultiAgent Systems (AAMAS'15): 357–364.

Bibliography

ANAC (2012). The Third International Automated Negotiating Agent Competition, http://anac2012.ecs.soton.ac.uk/results/final/.

Abdallah, S. and Lesser, V. R. (2005). Modeling task allocation using a decision theoretic model, *4th International Joint Conference on Autonomous Agents and Multiagent Systems (AAMAS)*, July 25–29, 2005, Utrecht, The Netherlands, pp. 719–726.

Abdallah, S. and Lesser, V. R. (2008). A multiagent reinforcement learning algorithm with non-linear dynamics, *Journal of Artificial Intelligence Research* **33**, 1, pp. 521–549.

Abdoos, M., Mozayani, N., and Bazzan, A. L. (2013). Holonic MultiAgent system for traffic signals control, *Engineering Applications of Artificial Intelligence* **26**, 5, pp. 1575–1587.

Ågotnes, T., van der Hoek, W., and Wooldridge, M. (2012). Conservative social laws, in *Proceedings of the 20th European Conference on Artificial Intelligence*, ECAI'12 (IOS Press), pp. 49–54.

Ågotnes, T. and Wooldridge, M. (2010). Optimal social laws, in *9th International Conference on Autonomous Agents and Multiagent Systems (AAMAS 2010)*, May 10–14, 2010, Toronto, Canada, Vol. 1–3, pp. 667–674.

Airiau, S., Sen, S., and Villatoro, D. (2014). Emergence of conventions through social learning, *Autonomous Agents and MultiAgent Systems* **28**, 5, pp. 779–804.

Albrecht, S. and Ramamoorthy, S. (2013a). A game-theoretic model and best-response learning method for *ad hoc* coordination in multiagent systems, Tech. rep., School of Informatics, The University of Edinburgh, United Kingdom.

Albrecht, S. V. and Ramamoorthy, S. (2013b). *Ad hoc* coordination in multiagent systems with applications to human-machine interaction, in *International Conference on Autonomous Agents and MultiAgent Systems (AAMAS)*, pp. 1415–1416, ISBN 978-1-4503-1993-5.

Allen, J. F. (1990). Maintaining knowledge about temporal intervals, *Readings in Qualitative Reasoning About Physical Systems* **26**, 11, pp. 361–372.

Alur, R., Henzinger, T. A., and Kupferman, O. (2002). Alternating-time temporal logic, *Journal of ACM* **49**, 5, pp. 672–713.

Amador, S., Okamoto, S., and Zivan, R. (2014). Dynamic MultiAgent task allocation with spatial and temporal constraints, in *Proceedings of 13th International Conference on Autonomous Agents and MultiAgent Systems*, pp. 1495–1496.

An, B., Lesser, V., Irwin, D., and Zink, M. (2010). Automated negotiation with decommitment for dynamic resource allocation in cloud computing, in *Proceedings of the 9th International Conference on Autonomous Agents and Multiagent Systems:* Vol. 1 (International Foundation for Autonomous Agents and Multiagent Systems), pp. 981–988.

Anagnostopoulos, A., Becchetti, L., Castillo, C., Gionis, A., and Leonardi, S. (2012). Online team formation in social networks, in *Proceedings of 21st International Conference on World Wide Web*, pp. 839–848.

Anari, N., Goel, G., and Nikzad, A. (2014). Mechanism design for crowdsourcing: An optimal 1-1/e competitive budget-feasible mechanism for large markets, in *Proceedings of 55th IEEE Symposium on Foundations of Computer Science*, pp. 266–275.

Andersen, S. and Jensen, F. (1989). Hugin: A shell for building belief universes for expert systems, in *International Joint Conference on Artificial Intelligence (IJCAI)*, pp. 332–337.

Angluin, D., Aspnes, J., Fischer, M. J., and Jiang, H. (2008). Self-stabilizing population protocols, *Acm Transactions on Autonomous & Adaptive Systems* **3**, 4, pp. 1–28.

Artikis, A., Kamara, L., Pitt, J., and Sergot, M. (2004). A protocol for resource sharing in norm-governed *ad hoc* networks, *Lecture Notes in Computer Science* **3476**, pp. 221–238.

Artikis, A., Sergot, M. J., and Pitt, J. V. (2009). Specifying norm-governed computational societies, *ACM Transaction on Computational Logic* **10**, 1, pp. 1:1–1:42.

Aspnes, J. and Ruppert, E. (2010). An introduction to population protocols, *Middleware for Network Eccentric & Mobile Applications* **93**, 93, p. 97.

Baarslag, T., Fujita, K., Gerding, E. H., Hindriks, K., Ito, T., Jennings, N. R., Jonker, C., Kraus, S., Lin, R., Robu, V., and Williams, C. R. (2013). Evaluating practical negotiating agents: Results and analysis of the 2011 international competition, *Artificial Intelligence* **198**, pp. 73–103.

Babaioff, M. and Blumrosen, L. (2008). Computationally-feasible truthful auctions for convex bundles, *Games & Economic Behavior* **63**, 2, pp. 588–620.

Bai, A., Wu, F., Zhang, Z., and Chen, X. (2014). Thompson sampling based Monte-Carlo planning in POMDPs, in *International Conference on Automated Planning and Scheduling (ICAPS)*, pp. 28–36.

Bai, H., Hsu, D., Kochenderfer, M., and Lee, W. (2011). Unmanned aircraft collision avoidance using continuous-state POMDPs, in *Robotics: Science and Systems (RSS)*, pp. 1–8.

Bailey, N. (1982). Spatial diffusion: An historical geography of epidemics in an island community: A. D. Cliff, P. Haggett, J. K. Ord and G. R. Versey cambridge: Cambridge university press, Cambridge geographical studies, 1981. pp. xi+238. 1950), *Journal of Historical Geography* **8**, 2, pp. 195–196.

Banerjee, B. and Peng, J. (2003). Adaptive policy gradient in multiagent learning, in *International Joint Conference on Autonomous Agents and Multiagent Systems*, pp. 686–692.

Barbulescu, L., Rubinstein, Z. B., Smith, S. F., and Zimmerman, T. L. (2010). Distributed coordination of mobile agent teams: The advantage of planning ahead, in *Proceedings of 9th International Conference on Autonomous Agents and Multiagent Systems*, pp. 1331–1338.

Bei, X., Chen, N., Gravin, N., and Lu, P. (2012). Budget feasible mechanism design: From prior-free to Bayesian, in *Proceedings of 44th ACM Symposium on Theory of Computing*, pp. 449–458.

Bellman, R. (1957). *Dynamic Programming* (Princeton University Press, Princeton, NJ, USA).

Benjacob, E. (1995). Novel type of phase transition in a system of self-driven particles, *Physical Review Letters* **75**, 6, p. 1226.

Berhault, M., Huang, H., Keskinocak, P., and Koenig, S. (2003). Robot exploration with combinatorial auctions, in *Proceedings of IEEE/RSJ International Conference on Intelligent Robots and Systems*, Vol. 2, pp. 1957–1962.

Bernstein, D. S., Givan, R., Immerman, N., and Zilberstein, S. (2002). The complexity of decentralized control of Markov decision processes, *Mathematics of Operations Research* **27**, 4, pp. 819–840.

Bianchi, R. A., Ribeiro, C. H., and Costa, A. H. R. (2007). Heuristic selection of actions in multiagent reinforcement learning. in *International Joint Conference on Artificial Intelligence*, pp. 690–695.

Binmore, K. (1992). Fun and Games: A Text on Game Theory (DC Heath and Company, Massachusetts, United States).

Biswas, A., Jain, S., Mandal, D., and Narahari, Y. (2015). A truthful budget feasible multi-armed bandit mechanism for crowdsourcing time critical tasks, in *Proceedings of 14th International Conference on Autonomous Agents and Multiagent Systems*, pp. 1101–1109.

Borodin, A., Filmus, Y., and Oren, J. (2010). Threshold models for competitive influence in social networks, in *International Conference on Internet and Network Economics*, pp. 539–550.

Boutilier, C. and Poole, D. (1996). Computing optimal policies for partially observable decision processes using compact representations, in *13th Conference on Artificial Intelligence (AAAI)*, pp. 1168–1175.

Bowling, M. (2004). Convergence and no-regret in multiagent learning, in *International Conference on Neural Information Processing Systems*, pp. 209–216.

Bowling, M. and Veloso, M. (2002). Multiagent learning using a variable learning rate, *Artificial Intelligence* **136**, 2, pp. 215–250.

Browne, C. B., Powley, E., Whitehouse, D., Lucas, S. M., Cowling, P. I., Rohlfshagen, P., Tavener, S., Perez, D., Samothrakis, S., and Colton, S. (2012). A survey of Monte-Carlo tree search methods, *IEEE Transactions on Computational Intelligence and AI in Games* **4**, 1, pp. 1–43.

Brucker, P. (2004). Scheduling algorithms (4th ed.), *Journal of the Operational Research Society* **47**, 8, pp. XII, 367.

Brzostowski, J. and Kowalczyk, R. (2006). Predicting partner's behaviour in agent negotiation, in *Proceedings of the Fifth International Joint Conference on Autonomous Agents and Multiagent Systems* (ACM, New York, NY, USA), pp. 355–361, ISBN 1-59593-303-4.

Buchfuhrer, D., Dughmi, S., Hu, F., Kleinberg, R., Mossel, E., Papadimitriou, C., Schapira, M., Singer, Y., and Umans, C. (2010). Inapproximability for vcg-based combinatorial auctions, in *Proceedings of 21st ACM-SIAM Symposium on Discrete Algorithms*, pp. 518–536.

Cai, Y., Daskalakis, C., and Weinberg, S. M. (2013a). Reducing revenue to welfare maximization: approximation algorithms and other generalizations, in *Proceedings of 24th ACM-SIAM Symposium on Discrete Algorithms*, pp. 578–595.

Cai, Y., Daskalakis, C., and Weinberg, S. M. (2013b). Understanding incentives: Mechanism design becomes algorithm design, in *Proceedings of 54th IEEE Symposium on Foundations of Computer Science*, pp. 618–627.

Campos, J., Esteva, M., López-Sánchez, M., Morales, J., and Salamó, M. (2011). Organisational adaptation of MultiAgent systems in a peer-to-peer scenario, *Computing* **91**, 2, pp. 169–215.

Cao, Z., Zhang, X. P., Cao, X. D., Yin, X., and Wang, D. (2009). A discussion on methodologies for research into complex systems, *Caai Transactions on Intelligent Systems*, **1**, p. 014.

Carbonneau, R., Kersten, G. E., and Vahidov, R. (2008). Predicting opponent's moves in electronic negotiations using neural networks, *Expert Systems with Applications* **34**, pp. 1266–1273.

Cassandra, A. (1998). *Exact and approximate algorithms for partially observable Markov decision processes*, Ph.D. Thesis, Brown University.

Cassandra, A., Littman, M., and Zhang, N. (1997). Incremental pruning: A simple, fast, exact method for partially observable Markov decision processes, in *Conference on Uncertainty in Artificial Intelligence (UAI)*, pp. 54–61.

Centola, D. (2010). The spread of behavior in an online social network experiment, *Science* **329**, 5996, p. 1194.

Chakraborty, T., Kearns, M., and Khanna, S. (2009). Network bargaining: Algorithms and structural results, in *Proceedings of 11th ACM Conference on Electronic Commerce*, pp. 159–168.

Chandra, P., Narahari, Y., Mandal, D., and Dey, P. (2015). Novel mechanisms for online crowdsourcing with unreliable, strategic agents, in *Proceedings of 29th AAAI Conference on Artificial Intelligence*, pp. 1256–1262.

Chandrasekaran, M., Doshi, P., Zeng, Y., and Chen, Y. (2014). Team behavior in interactive dynamic influence diagrams with applications to *ad hoc* teams (extended abstract), in *Autonomous Agents and MultiAgent Systems Conference (AAMAS)*, pp. 1559–1560.

Chapman, A. C., Micillo, R. A., Kota, R., and Jennings, N. R. (2009). Decentralised dynamic task allocation: A practical game: theoretic approach, in

Proceedings of 8th International Conference on Autonomous Agents and Multiagent Systems, pp. 915–922.

Chen, L., Qiu, X., Yang, Y., Gao, Z., and Qu, Z. (2012). The contract net based task allocation algorithm for wireless sensor network, in *Proceedings of IEEE Symposium on Computers and Communications*, pp. 600–604.

Chen, N., Gravin, N., and Lu, P. (2011). On the approximability of budget feasible mechanisms, in *Proceedings of 18th ACM-SIAM Symposium on Discrete Algorithms*, pp. 685–699.

Chen, N., Gravin, N., and Lu, P. (2014). Optimal competitive auctions, pp. 253–262.

Chen, S., Ammar, H. B., Tuyls, K., and Weiss, G. (2013a). Optimizing complex automated negotiation using sparse pseudo-input Gaussian processes, in *Proceedings of the 12th International Joint Conference on Autonomous Agents and MultiAgent Systems* (IFAAMAS), pp. 707–714.

Chen, S., Ammar, H. B., Tuyls, K., and Weiss, G. (2013b). Using conditional restricted boltzmann machine for highly competitive negotiation tasks, in *Proceedings of IJCAI'2013* (AAAI Press), pp. 69–75.

Chen, S., Hao, J., Weiss, G., Zhou, S., and Zhang, Z. (2015). Toward efficient agreements in real-time multilateral agent-based negotiations, in *2015 IEEE 27th International Conference on Tools with Artificial Intelligence (ICTAI)*, pp. 896–903, doi:10.1109/ICTAI.2015.130.

Chen, S. and Weiss, G. (2012). An efficient and adaptive approach to negotiation in complex environments, in *Proceedings of the 20th European Conference on Artificial Intelligence* (IOS Press), pp. 228–233.

Chen, S. and Weiss, G. (2013). An efficient automated negotiation strategy for complex environments, *Engineering Applications of Artificial Intelligence* **26**, 10, pp. 2613–2623.

Chen, S. and Weiss, G. (2014). An intelligent agent for bilateral negotiation with unknown opponents in continuous-time domains (in press), *ACM Transactions on Autonomous and Adaptive Systems*, **9**, 3, pp. 16:1–16:24.

Chen, Y., Hong, J., Liu, W., Godo, L., Sierra, C., and Loughlin, M. (2013c). Incorporating PGMs into a BDI architecture, in *16th International Conference on Principles and Practice of MultiAgent Systems (PRIMA)*, pp. 54–69.

Chevaleyre, Y., Endriss, U., and Maudet, N. (2007). Allocating goods on a graph to eliminate envy, in *Proceedings of Twenty-Second (AAAI) Conference on Artificial Intelligence*, July 22–26, Vancouver, British Columbia, Canada, pp. 700–705.

Chrisman, L. (1992). Reinforcement learning with perceptual aliasing: The perceptual distinctions approach, in *AAAI Conference on Artificial Intelligence (AAAI)*, pp. 183–188.

Clarke, E. H. (1971). Multipart pricing of public goods, *Public Choice* **11**, 11, pp. 17–33.

Claus, C. and Boutilier, C. (1998). The dynamics of reinforcement learning in cooperative multiagent systems, *AAAI/IAAI* **1998**, pp. 746–752.

Clearwater, S. H. (1996). *Market-Based Control: A Paradigm for Distributed Resource Allocation* (World Scientific, Singapore).

Coehoorn, R. M. and Jennings, N. R. (2004). Learning on opponent's preferences to make effective multi-issue negotiation trade-offs, in *Proceedings of the 6th International conference on Electronic commerce* (ACM, New York, NY, USA), pp. 59–68, ISBN 1-58113-930-6.

Conroy, R., Zeng, Y., Cavazza, M., and Chen, Y. (2015). Learning behaviors in agents systems with interactive dynamic influence diagrams, in *Proceedings of the 24th International Joint Conference on Artificial Intelligence*, pp. 39–45.

Conroy, R., Zeng, Y., Cavazza, M., Tang, J., and Pan, Y. (2016a). A value equivalence approach for solving interactive dynamic influence diagrams, in *Proceedings of the 2016 International Conference on Autonomous Agents & Multiagent Systems*, pp. 1162–1170.

Conroy, R., Zeng, Y., and Tang, J. (2016b). Approximating value equivalence in interactive dynamic influence diagrams using behavioral coverage, in *Proceedings of the 25th International Joint Conference on Artificial Intelligence*, pp. 201–207.

Criado, N., Argente, E., and Botti, V. (2011a). Open issues for normative MultiAgent systems, *AI Communications* **24**, 3, pp. 233–264.

Criado, N., Argente, E., Garrido, A., Gimeno, J. A., Igual, F., Botti, V., Noriega, P., and Giret, A. (2011b). Norm enforceability in electronic institutions? in *Coordination, Organizations, Institutions, and Norms in Agent Systems VI — COIN 2010 International Workshops, COIN@AAMAS 2010*, Toronto, Canada, May 2010, *COIN@MALLOW 2010*, Lyon, France, August 2010, Revised Selected Papers, pp. 250–267.

Cuesta, J. A., Gracia-Lázaro, C., Ferrer, A., Moreno, Y., and Sánchez, A. (2015). Reputation drives cooperative behaviour and network formation in human groups, *Scientific Reports* **5**, 7843, doi:10.1038/srep07843.

Dang, J. and Huhns, M. N. (2006). Concurrent multiple-issue negotiation for internet-based services, *IEEE Internet Computing* **10**, pp. 42–49.

Das, A., Gollapudi, S., and Munagala, K. (2014). Modeling opinion dynamics in social networks, in *Seventh ACM International Conference on Web Search and Data Mining, WSDM 2014* (New York, NY, USA), pp. 403–412.

Dash, R. K., Vytelingum, P., Rogers, A., David, E., and Jennings, N. R. (2007). Market-based task allocation mechanisms for limited-capacity suppliers, *IEEE Transactions on Systems Man & Cybernetics — Part A: Systems & Humans* **37**, 3, pp. 391–405.

Daubechies, I. (2006). *Ten Lectures on Wavelets* (Society for Industrial and Applied Mathematics), ISBN 9780898712742.

Davis, R. and Smith, R. G. (1983). Negotiation as a metaphor for distributed problem solving, *Artificial Intelligence* **20**, 1, pp. 63–109.

de Boor (1978). *A Practical Guide to Splines* (Springer-Verlag).

De Hauwere, Y.-M., Vrancx, P., and Nowé, A. (2010). Learning MultiAgent state space representations, in *Proceedings of the 9th International Conference on Autonomous Agents and Multiagent Systems:* Vol. 1 (International

Foundation for Autonomous Agents and Multiagent Systems), pp. 715–722.

Dechter, R., Meiri, I., and Pearl, J. (1991). Temporal constraint networks, *Artificial Intelligence* **49**, 1–3, pp. 61–95.

Degiovanni, R., Alrajeh, D., Aguirre, N., and Uchitel, S. (2014). Automated goal operationalisation based on interpolation and sat solving, in *International Conference on Software Engineering*, pp. 129–139.

Delgado, J. (2002). Emergence of social conventions in complex networks, *Artificial Intelligence*, **141**, 1/2, pp. 171–185.

Dias, M. B. (2004). Traderbots: A new paradigm for robust and efficient multi-robot coordination in dynamic environments, *Carnegie Mellon University*.

Dias, M. B., Zlot, R., Kalra, N., and Stentz, A. (2006). Market-based multirobot coordination: A survey and analysis, *Proceedings of the IEEE* **94**, 7, pp. 1257–1270.

Diener, E. (1984). Subjective well-being, *Psychological Bulletin* **95**, 3, pp. 542–575.

Diener, E., Sandvik, E., Seidlitz, L., and Diener, M. (1993). The relationship between income and subjective well-being: Relative or absolute? *Social Indicators Research* **28**, 3, pp. 195–223.

Doshi, P. and Gmytrasiewicz, P. J. (2006). On the difficulty of achieving equilibrium in interactive POMDPs, in *21st Conference on Artificial Intelligence (AAAI)*, pp. 1131–1136.

Doshi, P. and Perez, D. (2008). Generalized point based value iteration for interactive pomdps, in *Proceedings of the 23rd AAAI Conference on Artificial Intelligence, AAAI*, pp. 63–68.

Doshi, P. and Zeng, Y. (2009). Improved approximation of interactive dynamic influence diagrams using discriminative model updates, in *Autonomous Agents and MultiAgent Systems*, pp. 907–914.

Doshi, P., Zeng, Y., and Chen, Q. (2009). Graphical models for interactive pomdps: Representations and solutions, *Journal of Autonomous Agents and MultiAgent Systems (JAAMAS)* **18**, 3, pp. 376–416.

Duan, L., Dogru, M. K., Ozen, U., and Beck, J. (2012). A negotiation framework for linked combinatorial optimization problems, *Autonomous Agents and MultiAgent Systems* **25**, 1, pp. 158–182.

Eberling, M. and Büning, H. K. (2010). Self-adaptation strategies to favor cooperation, in *Proceedings of KES International Symposium on Agent and MultiAgent Systems: Technologies and Applications*, pp. 223–232.

Edalat, N., Xiao, W., Tham, C. K., and Keikha, E. (2009). A price-based adaptive task allocation for wireless sensor network, in *Proceedings of IEEE International Conference on Mobile Adhoc and Sensor Systems*, pp. 888–893.

Elmogy, A. M., Khamis, A. M., and Karray, F. O. (2009). Dynamic complex task allocation in multisensor surveillance systems, in *Proceedings of International Conference on Signals, Circuits and Systems*, pp. 1–6.

Faratin, P., Sierra, C., and Jennings, N. R. (1998). Negotiation decision functions for autonomous agents. *Robotics and Autonomous Systems* **24**, 4, pp. 159–182.

Faratin, P., Sierra, C., and Jennings, N. R. (2002). Using similarity criteria to make issue trade-offs in automated negotiations, *Artificial Intelligence* **142**, 2, pp. 205–237.

Farinelli, A., Scerri, P., and Tambe, M. (2003). Building large-scale robot systems: Distributed role assignment in dynamic, uncertain domains, in *Proceedings of AAMAS Workshop on Resources, Role and Task Allocation*, pp. 1–8.

Feng, Z. and Zilberstein, S. (2004). Region-based incremental pruning for POMDPs, in *Conference on Uncertainty in Artificial Intelligence (UAI)*, pp. 146–153.

Finnsson, H. and Björnsson, Y. (2008). Simulation-based approach to general game playing, in *AAAI Conference on Artificial Intelligence (AAAI)*, pp. 259–264.

FIPA (2001). *FIPA Contract Net Interaction Protocol Specification*, FIPA, http://www.fipa.org/specs/fipa00029/.

Franks, H., Griffiths, N., and Anand, S. S. (2014). Learning agent influence in mas with complex social networks, *Autonomous Agents and MultiAgent Systems* **28**, 5, pp. 836–866.

Franks, H., Griffiths, N., and Jhumka, A. (2013). Manipulating convention emergence using influencer agents, *Autonomous Agents and MultiAgent Systems* **26**, 3, pp. 315–353.

Friedman, E. J. and Parkes, D. C. (2003). Pricing wifi at starbucks: Issues in online mechanism design, in *Proceedings of 4th ACM Conference on Electronic Commerce*, pp. 240–241.

Fua, C. H. and Ge, S. S. (2005). Cobos: Cooperative backoff adaptive scheme for multirobot task allocation, *IEEE Transactions on Robotics* **21**, 6, pp. 1168–1178.

Gale, D. (1960). *The Theory of Linear Economic Models* (McGraw-Hill).

Gao, C. and Liu, J. (2017). Network-based modeling for characterizing human collective behaviors during extreme events, *IEEE Transactions on Systems, Man, and Cybernetics: Systems* **47**, 1, pp. 171–183.

Gao, Y. and Wei, W. (2006). Multi-robot autonomous cooperation integrated with immune based dynamic task allocation, in *Proceedings of International Conference on Intelligent Systems Design and Applications*, pp. 586–591.

Garg, S. K., Venugopal, S., Broberg, J., and Buyya, R. (2013). Double auction-inspired meta-scheduling of parallel applications on global grids, *Journal of Parallel & Distributed Computing* **73**, 4, pp. 450–464.

Gaston, M. E. and Desjardins, M. (2005). Agent-organized networks for dynamic team formation, in *Proceedings of 4th International Joint Conference on Autonomous Agents and Multiagent Systems*, pp. 230–237.

Gaston, M. E. and Desjardins, M. (2008). The effect of network structure on dynamic team formation in multiagent systems, *Computational Intelligence* **24**, 2, pp. 122–157.

Genesereth, M. and Thielscher, M. (2014). General game playing, *Synthesis Lectures on Artificial Intelligence and Machine Learning* **8**, 2, pp. 1–229.

Genesereth, M. R., Ginsberg, M. L., and Rosenschein, J. S. (1988). *Cooperation without Communication* (Morgan Kaufmann Publishers Inc.).

Genter, K. and Stone, P. (2016). Adding influencing agents to a flock, in *Proceedings of the 2016 International Conference on Autonomous Agents & Multiagent Systems*, Singapore, pp. 615–623.

Genter, K. and Stone, P. (2017). Agent behaviors for joining and leaving a flock, in *Proceedings of the 16th Conference on Autonomous Agents and MultiAgent Systems, AAMAS 2017*, São Paulo, Brazil, pp. 1553–1555.

Genter, K., Zhang, S., and Stone, P. (2015). Determining placements of influencing agents in a flock, in *Proceedings of the 2015 International Conference on Autonomous Agents and Multiagent Systems* (International Foundation for Autonomous Agents and Multiagent Systems), pp. 247–255.

Gerkey, B. P. and Matari, M. J. (2002). Sold!: Auction methods for multi-robot coordination, *IEEE Transactions on Robotics & Automation* **18**, 5, pp. 758–768.

Gerkey, B. P. and Mataric, M. J. (2004). A formal analysis and taxonomy of task allocation in multirobot systems, *International Journal of Robotics Research* **23**, 9, pp. 939–954.

Gmytrasiewicz, P. and Doshi, P. (2005). A framework for sequential planning in multiagent settings, *Journal of Artificial Intelligence Research (JAIR)* **24**, pp. 49–79.

Gode, D. K. and Sunder, S. (1993). Allocative efficiency of markets with zero-intelligence traders: Market as a partial substitute for individual rationality, *Journal of Political Economy* **101**, 1, pp. 119–137.

Godwin, M. F., Spry, S., and Hedrick, J. K. (2006). Distributed collaboration with limited communication using mission state estimates, in *Proceedings of American Control Conference*, p. 7.

Goldberg, A. V., Hartline, J. D., Karlin, A. R., Saks, M., and Wright, A. (2006). Competitive auctions, *Games & Economic Behavior* **55**, 2, pp. 242–269.

Goldberg, A. V., Hartline, J. D., and Wright, A. (2001). Competitive auctions and digital goods, in *Proceedings of 12th ACM-SIAM Symposium on Discrete Algorithms*, pp. 735–744.

Goldberg, D., Cicirello, V., Dias, M. B., Simmons, R., Smith, S., and Stentz, A. (2003). Market-based multi-robot planning in a distributed layered architecture, in *Proceedings of International Workshop on MultiRobot Systems*, **2**, pp. 27–38.

Gombolay, M., Wilcox, R. J., and Shah, J. A. (2013). Fast scheduling of multi-robot teams with temporospatial constraints. *Proccedings of Robotics: Science and Systems*, Berlin, Germany, doi:10.15607/RSS.2013.IX.049.

Griffiths, N. and Anand, S. S. (2012). The impact of social placement of non-learning agents on convention emergence, in *Proceedings of the 11th International Conference on Autonomous Agents and Multiagent Systems-Volume 3* (International Foundation for Autonomous Agents and Multiagent Systems), pp. 1367–1368.

Groves, T. (1973). Incentives in teams, *Econometrica* **41**, pp. 617–631.

Grubshtein, A., Gal-Oz, N., Grinshpoun, T., Meisels, A., and Zivan, R. (2010). Manipulating recommendation lists by global considerations, in *Proceedings of the 2nd International Conference on Agents and Artificial Intelligence (ICAART'10)*, pp. 135–142.

Grześ, M., Poupart, P., Yang, X., and Hoey, J. (2015). Energy efficient execution of POMDP policies, *IEEE Transactions on Cybernetics* **45**, 11, pp. 2484–2497.

Hao, J., Kang, E., Sun, J., and Jackson, D. (2016). Designing minimal effective normative systems with the help of lightweight formal methods, in *ACM Sigsoft International Symposium*, pp. 50–60.

Hao, J. and Leung, H. (2012). ABiNeS: An adaptive bilateral negotiating strategy over multiple items, in *Proceedings of WI/IAT'2012 (IEEE Computer Society)*, pp. 95–102.

Hao, J. and Leung, H.-f. (2013). The dynamics of reinforcement social learning in cooperative multiagent systems, in *Proceedings of 23rd International Joint Conference on Artificial Intelligence*, Vol. 13, pp. 184–190.

Hao, J., Sun, J., Chen, G., Wang, Z., Yu, C., and Ming, Z. (2017). Efficient and robust emergence of norms through heuristic collective learning, *ACM Transactions on Autonomous & Adaptive Systems* **12**, 4, pp. 1–20.

Hao, J., Sun, J., Huang, D., Cai, Y., and Yu, C. (2015). Heuristic collective learning for efficient and robust emergence of social norms, in *Proceedings of the 2015 International Conference on Autonomous Agents and Multiagent Systems*, pp. 1647–1648.

Hartline, J. (2010). Approximation in mechanism design, *American Economic Review* **102**, 3, pp. 330–336.

Hartline, J. and Karlin, A. (2007a). *Algorithmic Game Theory*, chap. Profit Maximization in Mechanism Design (Cambridge University Press), pp. 331–361.

Hartline, J. and Karlin, A. (2007b). *Algorithmic Game Theory*, chap. Online Mechanisms (Cambridge University Press), pp. 411–439.

Hauskrecht, M. (2000). Value-function approximations for partially observable Markov decision processes, in *Journal of Artificial Intelligence Research*, Vol. 13, pp. 33–94.

Heilporn, G., Raldine, Cordeau, J. F., and Laporte, G. (2010). The delivery man problem with time windows, *Discrete Optimization* **7**, 4, pp. 269–282.

Hendrikx, M. (2011). A survey of opponent models in automated negotiation, Tech. rep., Delft University of Technology, The Netherlands.

Hindriks, K. and Tykhonov, D. (2008). Opponent modelling in automated multi-issue negotiation using Bayesian learning, in *Proceedings of the 7th International Joint Conference on Autonomous Agents and Multiagent Systems* (ACM, Estoril, Portugal), pp. 331–338, ISBN 978-0-9817381-0-9.

Hoek, W. V. D., Roberts, M., and Wooldridge, M. (2007). Social laws in alternating time: Effectiveness, feasibility, and synthesis, *Synthese* **156**, 1, pp. 1–19.

Hoey, J., Poupart, P., Bertoldi, A. v., Craig, T., Boutilier, C., and Mihailidis, A. (2010). Automated handwashing assistance for persons with dementia using video and a partially observable Markov decision process, *Computer Vision and Image Understanding* **114**, 5, pp. 503–519.

Hornsby, K. (2003). Spatial diffusion: Conceptualizations and formalizations, *National Center for Geographic Information and Analysis and the Department of Spatial Information Science and Engineering*, University of Maine, p. 1. http://www.spatial.maine.edu/~khornsby/KHI21.pdf.

Hou, C. (2004). Predicting agents tactics in automated negotiation, in *IEEE / WIC / ACM International Conference on Intelligent Agent Technology*, Vol. 0 (IEEE Computer Society, Los Alamitos, CA, USA), ISBN 0-7695-2101-0, pp. 127–133.

Howard, R. A. and Matheson, J. E. (1984). Influence diagrams, in *Readings on the Principles and Applications of Decision Analysis*, pp. 721–762.

Hsiao, K., Kaelbling, L., and Lozano-Perez, T. (2007). Grasping POMDPs, in *IEEE International Conference on Robotics and Automation (ICRA)*, pp. 4685–4692.

Hsu, D., Lee, W., and Rong, N. (2007). What makes some POMDP problems easy to approximate, in *Advances in Neural Information Processing Systems (NIPS)*, pp. 689–696.

Hsu, T. S., Lee, J. C., Lopez, D. R., and Royce, W. A. (2001). Task allocation on a network of processors, *IEEE Transactions on Computers* **49**, 12, pp. 1339–1353.

Hu, Y., Gao, Y., and An, B. (2015a). Accelerating multiagent reinforcement learning by equilibrium transfer, *IEEE Transactions on Cybernetics* **45**, 7, pp. 1289–1302.

Hu, Y., Gao, Y., and An, B. (2015b). Learning in MultiAgent systems with sparse interactions by knowledge transfer and game abstraction, in *Proceedings of the 2015 International Conference on Autonomous Agents and Multiagent Systems* (International Foundation for Autonomous Agents and Multiagent Systems), pp. 753–761.

Hu, Y., Gao, Y., and An, B. (2015c). Multiagent reinforcement learning with unshared value functions, *IEEE Transactions on Cybernetics* **45**, 4, pp. 647–662.

Hunsberger, L. (2002). Algorithms for a temporal decoupling problem in MultiAgent planning, in *Proceedings of 18th AAAI Conference on Artificial Intelligence*, pp. 468–475.

Hussain, M., Kimiaghalam, B., Homaifar, A., Esterline, A., and Sayyarodsari, B. (2003). An evolutionary approach to capacitated resource distribution by a multiple-agent team, in *Proceedings of Genetic and Evolutionary Computation Conference*, pp. 657–668.

Inglehart, R., Foa, R., Peterson, C., and Welzel, C. (2008). Development, freedom, and rising happiness: A global perspective (1981–2007), *Perspectives on Psychological Science* **3**, 4, pp. 264–285.

Izakian, H., Abraham, A., and Ladani, B. T. (2010). An auction method for resource allocation in computational grids, *Future Generation Computer Systems* **26**, 2, pp. 228–235.

Jackson, D. (2002). Alloy: A lightweight object modelling notation, *ACM Transactions on Software Engineering & Methodology* **11**, 2, pp. 256–290.

Jadbabaie, A., Lin, J., and Morse, A. S. (2003). Coordination of groups of mobile autonomous agents using nearest neighbor rules, *IEEE Transactions on Automation Control* **48**, 6, pp. 988–1001.

Jain, S., Narayanaswamy, B., and Narahari, Y. (2014). A multiarmed bandit incentive mechanism for crowdsourcing demand response in smart grids, in *Proceeding of 28th AAAI Conference on Artificial Intelligence*, pp. 721–727.

Je, D. H., Choi, Y. H., and Seo, S. W. (2012). A heuristic task allocation methodology for designing the secure in-vehicle network, in *Proceedings of IEEE International Workshop on Vehicular Communications, Sensing, and Computing*, pp. 25–30.

Jennings, N. R. (2001). An agent-based approach for building complex software systems, *Communications of the ACM* **44**, 4, pp. 35–41.

Jennings, N. R., Faratin, P., Lomuscio, A. R., Parsons, S., Sierra, C., and Wooldridge, M. (2001). Automated negotiation: Prospects, methods, and challenges, *International Journal of Group Decision and Negotiation* **10**, 2, pp. 199–215.

Jennings, N. R., Moreau, L., Nicholson, D., Ramchurn, S., Roberts, S., Rodden, T., and Rogers, A. (2014). Human-agent collectives, *Communications of the ACM* **57**, 12, pp. 80–88.

Jeyabalan, V., Samraj, A., and Kiong, L. C. (2009). A market based approach for complex task allocation for wireless network based multirobot system, in *Proceedings of XXII International Symposium on Information, Communication and Automation Technologies*, pp. 1–5.

Jiang, J. and Xia, X. (2009). Prominence convergence in the collective synchronization of situated MultiAgents, *Information Processing Letters* **109**, 5, pp. 278–285.

Jiang, Y. (2008). Extracting social laws from unilateral binary constraint relation topologies in multiagent systems, *Expert Systems with Applications* **34**, 3, pp. 2004–2012.

Jiang, Y. (2009). Concurrent collective strategy diffusion of multiagents: The spatial model and case study, *IEEE Transactions Systems, Man, and Cybernetics, Part C* **39**, 4, pp. 448–458.

Jiang, Y. (2015). A survey of task allocation and load balancing in distributed systems, *IEEE Transactions on Parallel & Distributed Systems* **27**, 2, pp. 1–1.

Jiang, Y. and Hu, J. (2011). Favor-based decision: A novel approach to modeling the strategy diffusion in causal multiagent societies, *Expert Systems with Applications* **38**, 4, pp. 2974–2983.

Jiang, Y. and Huang, Z. (2012). The rich get richer: Preferential attachment in the task allocation of cooperative networked multiagent systems with resource caching, *IEEE Transactions on Systems Man and Cybernetics — Part A: Systems and Humans* **42**, 5, pp. 1040–1052.

Jiang, Y. and Ishida, T. (2007). Evolve individual agent strategies to global social law by hierarchical immediate diffusion, in *Massively MultiAgent*

Technology, *AAMAS Workshops, MMAS 2006, LSMAS 2006, and CCMMS 2007 Hakodate*, Japan, May 9, 2006, Honolulu, HI, USA, *Selected and Revised Papers*, pp. 80–91.

Jiang, Y. and Jiang, J. (2009). Contextual resource negotiation-based task allocation and load balancing in complex software systems, *IEEE Transactions on Parallel & Distributed Systems* **20**, 5, pp. 641–653.

Jiang, Y., Jiang, J., and Ishida, T. (2007). Agent coordination by trade-off between locally diffusion effects and socially structural influences, in *6th International Joint Conference on Autonomous Agents and Multiagent Systems (AAMAS 2007)*, Honolulu, Hawaii, USA, p. 77.

Jiang, Y. and Jiang, J. C. (2015). Diffusion in social networks: A multiagent perspective, *IEEE Transactions Systems, Man, and Cybernetics: Systems* **45**, 2, pp. 198–213.

Jiang, Y., Zhou, Y., and Li, Y. (2013a). Network layer-oriented task allocation for multiagent systems in undependable multiplex networks, in *Proceedings of 25TH IEEE International Conference on tools with Artificial Intelligence*, pp. 640–647.

Jiang, Y., Zhou, Y., and Li, Y. (2015). Reliable task allocation with load balancing in multiplex networks, *ACM Transactions on Autonomous & Adaptive Systems* **10**, 1, pp. 1–32.

Jiang, Y., Zhou, Y., and Wang, W. (2013b). Task allocation for undependable multiagent systems in social networks, *Parallel & Distributed Systems IEEE Transactions on* **24**, 8, pp. 1671–1681.

Jr, J. C. B. and Durfee, E. H. (2012). A distributed approach to summarizing spaces of multiagent schedules, in *Proceedings of Twenty-Sixth (AAAI) Conference on Artificial Intelligence*, July 22–26, Toronto, Ontario, Canada, pp. 1742–1748.

Jung, H. J., Park, Y., and Lease, M. (2014). Predicting next label quality: A time-series model of crowdwork, *AAAI Conference on Human Computation and Crowdsourcing*, pp. 1–9.

Kaelbling, L., Littman, M., and Cassandra, A. (1998). Planning and acting in partially observable stochastic domains, *Artificial Intelligence* **101**, 1–2, pp. 99–134.

Kafal, O. and Yolum, P. (2012). *Improving Self-organized Resource Allocation with Effective Communication* (Springer, Berlin, Heidelberg).

Kakade, S., Kearns, M., and Langford, J. (2003). Exploration in metric state spaces, in *International Conference on Machine Learning (ICML)*, pp. 206–312.

Kalai, E. and Lehrer, E. (1993). Rational learning leads to Nash equilibrium, *Econometrica*, **61**, 5, pp. 1019–1045.

Kalai, E. and Smorodinsky, M. (1975). Other solutions to Nash's bargaining problem, *Econometrica: Journal of the Econometric Society*, **43**, 3, pp. 513–518.

Kalra, N. and Martinoli, A. (2006). Comparative study of market-based and threshold-based task allocation, *Distributed Autonomous Robotic Systems* **7**, pp. 91–101.

Kapetanakis, S. and Kudenko, D. (2002a). Reinforcement learning of coordination in cooperative MultiAgent systems, *Lecture Notes in Computer Science* **3**, 4, pp. 1258–1259.

Kapetanakis, S. and Kudenko, D. (2002b). Reinforcement learning of coordination in cooperative multiagent systems, in *Proceedings of the 18th National Conference on Artificial Intelligence (AAAI)*, pp. 326–331.

Kapetanakis, S. and Kudenko, D. (2005). Reinforcement learning of coordination in heterogeneous cooperative multiagent systems, in *Adaptive Agents and MultiAgent Systems II* (Springer), pp. 119–131.

Kargar, M. and An, A. (2011). Discovering top-k teams of experts with/without a leader in social networks, in *Proceedings of 20th ACM Conference on Information and Knowledge Management*, pp. 985–994.

Kennes, J. (2006). *Competitive Auctions: Theory and Application* (Elsevier Science & Technology).

Khamis, A. M., Elmogy, A. M., and Karray, F. O. (2011). Complex task allocation in mobile surveillance systems, *Journal of Intelligent & Robotic Systems* **64**, 1, pp. 33–55.

Kleinberg, J. (2008). Balanced outcomes in social exchange networks, in *Proceedings of 40th ACM Symposium on Theory of Computing*, pp. 295–304.

Kobayashi, S., Kyoshima, K., Olschowka, J. A., and Jacobowitz, D. M. (2005). Solving the auction-based task allocation problem in an open environment. in *Proceedings of 20th AAAI Conference on Artificial Intelligence*, pp. 164–169.

Kochenderfer, M. J. (2015). *Decision Making Under Uncertainty: Theory and Application* (MIT Press, Cambridge, Massachusetts, USA).

Kocsis, L. and Szepesvári, C. (2006). Bandit based Monte-Carlo planning, in *European Conference on Machine Learning (ECML)*, pp. 282–293.

Koes, M., Nourbakhsh, I., and Sycara, K. (2005). Heterogeneous multirobot coordination with spatial and temporal constraints, in *Proceedings of 20th AAAI Conference on Artificial Intelligence and 17th Innovative Applications of Artificial Intelligence Conference*, pp. 1292–1297.

Kok, J. R., Vlassis, N., *et al.* (2004). Sparse tabular multiagent q-learning, in *Annual Machine Learning Conference of Belgium and the Netherlands*, pp. 65–71.

Korsah, G. A., Stentz, A., and Dias, M. B. (2013). A comprehensive taxonomy for multi-robot task allocation, *International Journal of Robotics Research* **32**, 12, pp. 1495–1512.

Kota, R., Gibbins, N., and Jennings, N. R. (2009). Self-organising agent organisations, in *Proceedings of 8th International Conference on Autonomous Agents and Multiagent Systems*, pp. 797–804.

Koutsoupias, E. and Papadimitriou, C. H. (1999). Worst-case equilibria, in *STACS 99, 16th Annual Symposium on Theoretical Aspects of Computer Science*, Trier, Germany, March 4–6, 1999, pp. 404–413.

Krothapalli, N. and Deshmukh, A. V. (2004). Dynamic allocation of communicating tasks in computational grids, *IIE Transactions*, 11, pp. 1037–1053.

Kuhn, F., Lynch, N., and Oshman, R. (2010). Distributed computation in dynamic networks, in *Proceedings of the ACM Symposium on Theory of Computing*, pp. 513–522.

Kurniawati, H., Hsu, D., and Lee, W. (2008). SARSOP: Efficient point-based POMDP planning by approximating optimally reachable belief spaces, in *Robotics: Science and Systems (RSS)*.

Lagoudakis, M. G., Markakis, E., Kempe, D., Keskinocak, P., Kleywegt, A. J., Koenig, S., Tovey, C. A., Meyerson, A., and Jain, S. (2005). Auction-based multi-robot routing, in *Robotics: Science and Systems*, pp. 343–350.

Lai, G., Li, C., Sycara, K., and Giampapa, J. (2004). *Literature review on multi-attribute negotiations*, Tech. rep., Robotics Institute, Carnegie Mellon University.

Landn, D., Heintz, F., and Doherty, P. (2010). *Complex Task Allocation in Mixed-Initiative Delegation: A UAV Case Study* (Springer Berlin Heidelberg).

Lappas, T., Liu, K., and Terzi, E. (2009). Finding a team of experts in social networks, in *Proceedings of 15th ACM International Conference on Knowledge Discovery and Data Mining*, pp. 467–476.

Lau, H. C. and Zhang, L. (2003). Task allocation via MultiAgent coalition formation: Taxonomy, algorithms and complexity, in *Proceedings of 15th IEEE International Conference on Tools with Artificial Intelligence*, pp. 346–350.

Lau, R. Y., Li, Y., Song, D., and Kwok, R. C. W. (2008). Knowledge discovery for adaptive negotiation agents in e-marketplaces, *Decision Support Systems* **45**, 2, pp. 310–323.

Lavi, R. and Nisan, N. (2004). Competitive analysis of incentive compatible on-line auctions, *Theoretical Computer Science* **310**, 1, pp. 159–180.

Lavi, R. and Nisan, N. (2005). Online ascending auctions for gradually expiring items, in *Proceedings of 16th ACM-SIAM Symposium on Discrete Algorithms*, pp. 1146–1155.

Lemaire, T., Alami, R., and Lacroix, S. (2004). A distributed tasks allocation scheme in multi-UAV context, in *Proceedings of IEEE International Conference on Robotics and Automation*, Vol. 4, pp. 3622–3627.

Lerman, K., Jones, C., Galstyan, A., Matar, and C, M. J. (2006). Analysis of dynamic task allocation in multirobot systems, *International Journal of Robotics Research* **25**, 3, pp. 225–241.

Li, X. and Yao, A. C. (2013). On revenue maximization for selling multiple independently distributed items, *Proceedings of the National Academy of Sciences* **110**, 28, pp. 11232–11237.

Li, X., Zhang, C., Hao, J., Tuyls, K., Chen, S., and Feng, Z. (2016). Socially-aware multiagent learning: Towards socially optimal outcomes. in *ECAI*, pp. 533–541.

Li, Y., Gao, Z., Yang, Y., and Guan, Z. (2010). A cluster-based negotiation model for task allocation in wireless sensor network, in *Proceedings of International Conference on Network and Service Management*, pp. 112–117.

Lin, L., Lei, W., Zheng, Z., and Sun, Z. (2004a). A learning market based layered multi-robot architecture, in *Proceedings of IEEE International Conference on Robotics and Automation*, Vol. 4, pp. 3417–3422.

Lin, R. and Kraus, S. (2010). Can automated agents proficiently negotiate with humans? *Communications of the ACM* **53**, 1, pp. 78–88.

Lin, R., Kraus, S., Wilkenfeld, J., and Barry, J. (2008). Negotiating with bounded rational agents in environments with incomplete information using an automated agent, *Artificial Intelligence* **172**, pp. 823–851.

Lin, Z., Broucke, M. E., and Francis, B. A. (2004b). Local control strategies for groups of mobile autonomous agents, *IEEE Transactions on Automation Control* **49**, 4, pp. 622–629.

Littman, M. (1996). *Algorithms for Sequential Decision Making*, Ph.D. Thesis, Brown University.

Littman, M., Cassandra, A., and Kaelbling, L. (1995). Learning policies for partially observable environments: Scaling up, in *International Conference on Machine Learning (ICML)*, pp. 362–370.

Liu, J., Jin, X., and Tsui, K. C. (2005a). Autonomy oriented computing, *Multiagent Systems Artificial Societies & Simulated Organizations* **12**, 6, pp. 879–902.

Liu, J., Jin, X., and Wang, Y. (2005b). Agent-based load balancing on homogeneous minigrids: Macroscopic modeling and characterization, *IEEE Transactions on Parallel & Distributed Systems* **16**, 7, pp. 586–598.

Lomuscio, A. R., Wooldridge, M., and Jennings, N. R. (2003). A classification scheme for negotiation in electronic commerce, *Group Decision and Negotiation* **12**, 1, pp. 31–56.

Lopes, F., Wooldridge, M., and Novais, A. (2008). Negotiation among autonomous computational agents: Principles, analysis and challenges, *Artificial Intelligence Review* **29**, pp. 1–44.

Low, K. H., Leow, W. K., and Ang, M. H. (2004). Task allocation via self-organizing swarm coalitions in distributed mobile sensor network, in *Proceedings of 19th AAAI Conference on Artificial Intelligence*, pp. 28–33.

Low, K. H., Leow, W. K., and Ang, V. M. H. (2006). Autonomic mobile sensor network with self-coordinated task allocation and execution, *IEEE Transactions on Systems Man & Cybernetics — Part C Applications & Reviews* **36**, 3, pp. 315–327.

Luo, J., Yin, H., Li, B., and Wu, C. (2011). Path planning for automated guided vehicles system via interactive dynamic influence diagrams with communication, in *9th IEEE International Conference on Control and Automation (ICCA)*, pp. 755–759.

MacAlpine, P., Genter, K. L., Barrett, S., and Stone, P. (2014). The robocup 2013 drop-in player challenges: A testbed for *ad hoc* teamwork, in *International conference on Autonomous Agents and MultiAgent Systems, AAMAS '14*, Paris, France, pp. 1461–1462.

Macarthur, K. S., Stranders, R., Ramchurn, S. D., and Jennings, N. R. (2011). A distributed anytime algorithm for dynamic task allocation in MultiAgent systems, in *Proceedings of 25th AAAI Conference on Artificial Intelligence*, pp. 701–706.

Madani, O., Hanks, S., and Condon, A. (1999). On the undecidability of probabilistic planning and infinite-horizon partially observable Markov decision problems, in *AAAI Conference on Artificial Intelligence (AAAI)*, pp. 541–548.

Mahmoud, S., Griffiths, N., Keppens, J., and Luck, M. (2012). Norm emergence: Overcoming hub effects in scale free networks, in *Proceedings of the AAMAS 2012 Workshop on Coordination, Organizations, Institutions and Norms*, pp. 136–150.

Majumder, A., Datta, S., and Naidu, K. V. M. (2012). Capacitated team formation problem on social networks, in *Proceedings of 18th ACM International Conference on Knowledge Discovery and Data Mining*, pp. 1005–1013.

Marchant, J. and Griffiths, N. (2015). Manipulating conventions in a particle-based topology, in *Coordination, Organizations, Institutions, and Normes in Agent Systems XI — COIN 2015 International Workshops, COIN@AAMAS*, Istanbul, Turkey, *COIN@IJCAI*, Buenos Aires, Argentina, *Revised Selected Papers*, pp. 242–261.

Marchant, J., Griffiths, N., and Leeke, M. (2014a). Destabilising conventions: Characterising the cost, in *8th IEEE International Conference on Self-Adaptive and Self-Organizing Systems, SASO 2014*, London, United Kingdom, 2014, pp. 139–144.

Marchant, J., Griffiths, N., Leeke, M., and Franks, H. (2014b). Destabilising conventions using temporary interventions, in *Coordination, Organizations, Institutions, and Norms in Agent Systems X — COIN 2014 International Workshops, COIN@AAMAS*, Paris, France, *COIN@PRICAI*, Gold Coast, QLD, Australia, *Revised Selected Papers*, pp. 148–163.

Marecki, J., Gupta, T., Varakantham, P., Tambe, M., and Yokoo, M. (2008). Not all agents are equal: Scaling up distributed POMDPs for agent networks, in *International Joint Conference on Autonomous Agents and MultiAgent Systems (AAMAS)*, pp. 485–492.

Mashayekhy, L., Nejad, M. M., Grosu, D., and Vasilakos, A. (2016). An online mechanism for resource allocation and pricing in clouds, *IEEE Transactions on Computers* **65**, 4, pp. 1172–1184.

Mi, Z., Yang, Y., Ma, H., and Wang, D. (2014). Connectivity preserving task allocation in mobile robotic sensor network, in *Proceedings of IEEE International Conference on Communications*, pp. 136–141.

Mihaylov, M., Tuyls, K., and Nowé, A. (2014). A decentralized approach for convention emergence in MultiAgent systems, *Autonomous Agents and MultiAgent Systems* **28**, 5, pp. 749–778.

Mnih, V., Larochelle, H., and Hinton, G. E. (2012). Conditional restricted Boltzmann machines for structured output prediction, *CoRR* **abs/1202.3748**.

Monahan, G. E. (1982). A survey of partially observable Markov decision processes: Theory, models, and algorithms, *Management Science* **28**, 1, pp. 1–16.

Mor, Y., Goldman, C. V., and Rosenschein, J. S. (1996). Learn your opponent's strategy (in polynomial time)! in *Proceedings of IJCAI-95 Workshop on Adaptation and Learning in Multiagent Systems* (Springer-Verlag), pp. 164–176.

Morales, J., Lopez-Sanchez, M., Rodriguez-Aguilar, J. A., Wooldridge, M., and Vasconcelos, W. (2013a). Automated synthesis of normative systems, in *International Conference on Autonomous Agents and MultiAgent Systems*, pp. 109–116.

Morales, J., López-Sánchez, M., Rodríguez-Aguilar, J. A., Wooldridge, M., and Vasconcelos, W. W. (2013b). Automated synthesis of normative systems, in *International conference on Autonomous Agents and MultiAgent Systems*, *AAMAS '13*, Saint Paul, MN, USA, pp. 483–490.

Morales, J., Lpez-Snchez, M., and Esteva, M. (2011). Using experience to generate new regulations, in *IJCAI 2011, Proceedings of the International Joint Conference on Artificial Intelligence*, Barcelona, Catalonia, Spain, July, pp. 307–312.

Moshtagh, N. and Jadbabaie, A. (2007). Distributed geodesic control laws for flocking of nonholonomic agents, *IEEE Transactions Automatation Control* **52**, 4, pp. 681–686.

Mukherjee, P., Sen, S., and Airiau, S. (2008). Norm emergence under constrained interactions in diverse societies, in *Proceedings of the 7th International Joint Conference on Autonomous Agents and Multiagent Systems*, Vol. 2 (International Foundation for Autonomous Agents and Multiagent Systems), pp. 779–786.

Muthoo, A. (1999). *Bargaining Theory with Applications* (Cambridge University Press).

Myerson, R. B. (1981a). Optimal auction design, *Mathematics of Operations Research* **6**, 1, pp. 58–73.

Myerson, R. B. (1981b). Utilitarianism, egalitarianism, and the timing effect in social choice problems, *Econometrica: Journal of the Econometric Society*, **49**, 4, pp. 883–897.

Nanjanath, M. and Gini, M. (2006). Dynamic task allocation for robots via auctions, in *Proceedings of IEEE International Conference on Robotics and Automation*, pp. 2781–2786.

Nash, J. (1953). Two-person cooperative games, *Econometrica* **21**, 1, pp. 128–140, http://www.jstor.org/stable/1906951.

Nash, J. F. *et al.* (1950). Equilibrium points in n-person games, *Proceedings of the National Academy of Sciences* **36**, 1, pp. 48–49.

Neely, M. J. (2010). *Stochastic Network Optimization with Application to Communication and Queueing Systems* (Morgan and Claypool Publishers).

Nisan, N. and Ronen, A. (2004). Computationally feasible vcg mechanisms, *Journal of Artificial Intelligence Research* **29**, 6, pp. 242–252.

Nisan, N. and Ronen, A. (1999). Algorithmic mechanism design, in *Proceedings of 31st ACM Symposium on Theory of Computing*, pp. 129–140.

Nisan, N. and Ronen, A. (2001). Algorithmic mechanism design, *Games & Economic Behavior* **35**, 12, pp. 166–196.

Nunes, E. and Gini, M. (2015). Multi-robot auctions for allocation of tasks with temporal constraints, in *Proceedings of 29th AAAI Conference on Artificial Intelligence*, pp. 2110–2216.

Nunes, E., Manner, M., Mitiche, H., and Gini, M. (2017). A taxonomy for task allocation problems with temporal and ordering constraints, *Robotics and Autonomous Systems*, **90**, pp. 55–70.

Oh, J. and Smith, S. F. (2008). A few good agents: MultiAgent social learning, in *7th International Joint Conference on Autonomous Agents and Multiagent Systems (AAMAS 2008)*, Estoril, Portugal, May 12–16, 2008, Vol. 1, pp. 339–346.

Olfati-Saber, R. (2006). Flocking for MultiAgent dynamic systems: algorithms and theory, *IEEE Transactions on Automatatic Control* **51**, 3, pp. 401–420.

Olfati-Saber, R., Fax, J. A., and Murray, R. M. (2007). Consensus and cooperation in networked MultiAgent systems, in *Proceedings of the IEEE* **95**, 1, pp. 215–233.

Osborne, M. and Rubinstein, A. (1994a). *A Course in Game Theory* (MIT Press).

Osborne, M. J. and Rubinstein, A. (1994b). *A Course in Game Theory* (MIT press).

Oshrat, Y., Lin, R., and Kraus, S. (2009). Facing the challenge of human-agent negotiations via effective general opponent modeling, in *Proceedings of The 8th International Conference on Autonomous Agents and Multiagent Systems*, Vol. 1 (International Foundation for Autonomous Agents and Multiagent Systems), pp. 377–384.

Pacheco, J. M., Traulsen, A., and Nowak, M. A. (2006). Coevolution of strategy and structure in complex networks with dynamical linking, *Physical Review Letters* **97**, 25, p. 258103.

Page, A. J., Keane, T. M., and Naughton, T. J. (2010). Multi-heuristic dynamic task allocation using genetic algorithms in a heterogeneous distributed system, *Journal of Parallel & Distributed Computing* **70**, 7, pp. 758–766.

Pan, S., Larson, K., Bradshaw, J., and Law, E. (2016). Dynamic task allocation algorithm for hiring workers that learn, in *Proceedings of 25th International Joint Conference on Artificial Intelligence*, pp. 3825–3831.

Papadimitriou, C. and Tsitsiklis, J. (1987). The complexity of Markov decision processes, *Mathematics of Operations Research* **12**, 3, pp. 441–450.

Pineau, J., Gordon, G., and Thrun, S. (2003). Point-based value iteration: An anytime algorithm for POMDPs, in *International Joint Conference on Artificial Intelligence (IJCAI)*, pp. 1025–1032.

Pineau, J., Gordon, G., and Thrun, S. (2006a). Anytime point-based approximations for large POMDPs, *Journal of Artificial Intelligence Research* **27**, pp. 335–380.

Pineau, J., Gordon, G., and Thrun, S. (2006b). Anytime point-based value iteration for large pomdps, *Journal of Artificial Intelligence Research* **27**, pp. 335–380.

Pizzocaro, D. and Preece, A. (2009). Towards a taxonomy of task allocation in sensor networks, in *INFOCOM Workshops*, pp. 1–2.

Ponda, S. S., Johnson, L. B., Kopeikin, A. N., and Choi, H. L. (2012). Distributed planning strategies to ensure network connectivity for dynamic heterogeneous teams, *IEEE Journal on Selected Areas in Communications* **30**, 5, pp. 861–869.

Pongpunwattana, A., Rysdyk, R., Vagners, J., and Rathbun, D. (2006). Market-based co-evolution planning for multiple autonomous vehicles, in *Proceedings of AIAA Unmanned Unlimited Systems, Technologies & Operations Conference*, pp. 2003–6524.

Ponka, I. (2009). *Commitment models and concurrent bilateral negotiation strategies in dynamic service markets*, Ph.D. Thesis, University of Southampton, School of Electronics and Computer Science.

Poupart, P. (2005). *Exploiting structure to efficiently solve large scale partially observable Markov decision processes*, Ph.D. Thesis, University of Toronto.

Poupart, P., Kim, K., and Kim, D. (2011). Closing the gap: Improved bounds on optimal POMDP solutions, in *International Conference on Automated Planning and Scheduling (ICAPS)*, pp. 194–201.

Ragone, A., Noia, T., Sciascio, E., and Donini, F. (2008). Logic-based automated multi-issue bilateral negotiation in peer-to-peer e-marketplaces, *Autonomous Agents and MultiAgent Systems* **16**, 3, pp. 249–270.

Rahmandad, H. and Sterman, J. (2008). Heterogeneity and network structure in the dynamics of diffusion: Comparing agent-based and differential equation models, *Management Science* **54**, 5, pp. 998–1014.

Raiffa, H. (1982). *The Art and Science of Negotiation* (Harvard University Press Cambridge, Mass), ISBN 9780674048133.

Rangapuram, S. S., Bhler, T., and Hein, M. (2015). Towards realistic team formation in social networks based on densest subgraphs, in *Proceedings of International Conference on World Wide Web*, pp. 1077–1088.

Reynolds, C. W. (1987). Flocks, herds and schools: A distributed behavioral model, in *Conference on Computer Graphics and Interactive Techniques*, pp. 25–34.

Ross, S. and Chaib-Draa, B. (2007). AEMS: An anytime online search algorithm for approximate policy refinement in large POMDPs, in *International Joint Conference on Artificial Intelligence (IJCAI)*, pp. 2592–2598.

Ross, S., Pineau, J., Paquet, S., and Chaib-Draa, B. (2008). Online planning algorithms for POMDPs, *Journal of Artificial Intelligence Research (JAIR)* **32**, 1, pp. 663–704.

Russell, S. and Norvig, P. (2010). *Artificial Intelligence: A Modern Approach (Third Edition)* (Prentice Hall).

Sacks, D. W., Stevenson, B., and Wolfers, J. (2010). Subjective well-being, income, economic development and growth, *The National Bureau of Economic Research Working Paper*, 16441, pp. 1–53.

Saha, S., Biswas, A., and Sen, S. (2005). Modeling opponent decision in repeated one-shot negotiations, in *Proceedings of the Fourth international joint conference on Autonomous agents and multiagent systems* (ACM, New York, NY, USA), pp. 397–403, ISBN 1-59593-093-0.

Sandholm, T. (1993). An implementation of the contract net protocol based on marginal cost calculations, in *Proceedings of 11th AAAI Conference on Artificial Intelligence*, pp. 256–262.

Sandholm, T., Larson, K., Andersson, M., Shehory, O., and Tohm, F. (1999). Coalition structure generation with worst case guarantees, *Artificial Intelligence* **111**, 1–2, pp. 209–238.

Sandholm, T. W. (1998). Contract types for satisfying task allocation, in *Proceedings of AAAI Spring Symposium Satisficing Models*, pp. 68–75.

Sandholm, T. W. and Lesser, V. R. (1996). Advantages of a leveled commitment contracting protocol, in *Proceedings of 13th AAAI Conference on Artificial Intelligence*, pp. 126–133.

Sariel, S., Balch, T. R., and Erdogan, N. (2008). Robust multi-robot cooperation through dynamic task allocation and precaution routines. in *Proceedings of 3rd International Conference on Informatics in Control, Automation and Robotics*, pp. 196–201.

Satia, J. K. and Lave, R. E. (1973). Markovian decision processes with probabilistic observation of state, *Management Science* **20**, 1, pp. 1–13.

Savarimuthu, B. T. R. (2011). Norm learning in Multi-agent societies, *Information Science Discussion Papers Series*, 2011/05, University of Otago. http://hdl.handle.net/10523/1690.

Savarimuthu, B. T. R., Arulanandam, R., and Purvis, M. (2011). Aspects of active norm learning and the effect of lying on norm emergence in agent societies, in *Agents in Principle, Agents in Practice* (Springer), pp. 36–50.

Savarimuthu, B. T. R. and Cranefield, S. (2011). Norm creation, spreading and emergence: A survey of simulation models of norms in MultiAgent systems, *Multiagent and Grid Systems* **7**, 1, pp. 21–54.

Scerri, P., Farinelli, A., Okamoto, S., and Tambe, M. (2005). Allocating tasks in extreme teams, in *Proceedings of 4th International Joint Conference on Autonomous Agents and Multiagent Systems*, pp. 727–734.

Schneider, J., Apfelbaum, D., Bagnell, D., and Simmons, R. (2005). Learning opportunity costs in multi-robot market based planners, in *Proceedings of IEEE International Conference on Robotics and Automation*, pp. 1151–1156.

Schoenig, A. and Pagnucco, M. (2010). Evaluating sequential single-item auctions for dynamic task allocation, in *Proceedings of Australasian Joint Conference on Advances in Artificial Intelligence*, pp. 506–515.

Schwartenbeck, P., FitzGerald, T. H. B., Mathys, C., Dolan, R., Kronbichler, M., and Friston, K. (2015). Evidence for surprise minimization over value maximization in choice behavior, *Scientific Reports* **5**, 16575, doi:10.1038/srep16575.

Sen, O. and Sen, S. (2010). Effects of social network topology and options on norm emergence, in *Coordination, Organizations, Institutions and Norms in Agent Systems V* (Springer), pp. 211–222.

Sen, S. (2013). A comprehensive approach to trust management, in *Proceedings of the 12th International Conference on Autonomous Agents and MultiAgent Systems (AAMAS'13)*, pp. 797–800.

Sen, S. and Airiau, S. (2007). Emergence of norms through social learning. in *International Joint Conference on Artificial Intelligence*, Vol. 1507, p. 1512.

Service, T. C. and Adams, J. A. (2011). Coalition formation for task allocation: Theory and algorithms, *Autonomous Agents and MultiAgent Systems* **22**, 2, pp. 225–248.

Seuken, S. and Zilberstein, S. (2007). Memory bounded dynamic programming for decentralized POMDPs, in *International Joint Conference on Artificial Intelligence (IJCAI)*, pp. 2009–2015.

Shachter, R. D. (1986). Evaluating influence diagrams, *Operations Research* **34**, 6, pp. 871–882.

Shani, G., Brafman, R., and Shimony, S. (2007). Forward search value iteration for POMDPs, in *International Joint Conference on Artificial Intelligence (IJCAI)*, pp. 2619–2624.

Shani, G. and Meek, C. (2009). Improving existing fault recovery policies, in *Advances in Neural Information Processing Systems (NIPS)*, pp. 1642–1650.

Shani, G., Pineau, J., and Kaplow, R. (2013). A survey of point-based POMDP solvers, *Journal of Autonomous Agents and MultiAgent Systems* **27**, 1, pp. 1–51.

Shehory, O. (1999). *A Scalable Agent Location Mechanism* (Springer, Berlin, Heidelberg).

Shehory, O. and Kraus, S. (1995a). Task allocation via coalition formation among autonomous agents, in *Proceedings of International Joint Conference on Artificial Intelligence*, pp. 655–661.

Shehory, O. and Kraus, S. (1995b). Task allocation via coalition formation among autonomous agents, in *Proceedings of International Joint Conference on Artificial Intelligence*, pp. 655–661.

Shehory, O. and Kraus, S. (1998). Methods for task allocation via agent coalition formation, *Artificial Intelligence* **101**, 1–2, pp. 165–200.

Shoham, Y. and Leyton-Brown, K. (2009). *Multiagent Systems: Algorithmic, Game-theoretic, and Logical Foundations* (Cambridge University Press).

Shoham, Y. and Tennenholtz, M. (1992). On the synthesis of useful social laws for artificial agent societies, in *10th National Conference on Artificial Intelligence*, pp. 276–281.

Shoham, Y. and Tennenholtz, M. (1995). On social laws for artificial agent societies: off-line design, *Artificial Intelligence* **73**, 1, pp. 231–252.

Shoham, Y. and Tennenholtz, M. (1997). On the emergence of social conventions: modeling, analysis, and simulations, *Artificial Intelligence* **94**, 1, pp. 139–166.

Silver, D., Huang, A., Maddison, C. J., Guez, A., Sifre, L., van den Driessche, G., Schrittwieser, J., Antonoglou, I., Panneershelvam, V., Lanctot, M., Dieleman, S., Grewe, D., Nham, J., Kalchbrenner, N., Sutskever, I., Lillicrap, T., Leach, M., Kavukcuoglu, K., Graepel, T., and Hassabis, D. (2016). Mastering the game of Go with deep neural networks and tree search, *Nature* **529**, pp. 484–489.

Silver, D. and Veness, J. (2010). Monte-Carlo planning in large POMDPs, in *Advances in Neural Information Processing Systems (NIPS)*, pp. 2164–2172.

Singer, Y. (2010). Budget feasible mechanisms, in *Proceedings of 51st IEEE Symposium on Foundations of Computer Science*, pp. 765–774.

Singh, S., Kearns, M., and Mansour, Y. (2000). Nash convergence of gradient dynamics in general-sum games, in *Proceedings of the 16th conference on Uncertainty in artificial intelligence*, pp. 541–548.

Smith, R. G. (1980). The contract net protocol: High-level communication and control in a distributed problem solver, *IEEE Transactions on Computers* **c-29**, 12, pp. 1104–1113.

Smith, T. (2007). *Probabilistic planning for robotic exploration*, Ph.D. Thesis, Carnegie Mellon University.

Smith, T. and Simmons, R. (2004). Heuristic search value iteration for POMDPs, in *Conference on Uncertainty in Artificial Intelligence (UAI)*, pp. 520–527.

Smith, T. and Simmons, R. (2005). Point-based POMDP algorithms: Improved analysis and implementation, in *Conference on Uncertainty in Artificial Intelligence (UAI)*, pp. 542–547.

Somani, A., Ye, N., Hsu, D., and Lee, W. (2013). DESPOT: Online POMDP planning with regularization, in *Advances in Neural Information Processing Systems (NIPS)*, pp. 1772–1780.

Sondik, E. (1971). *The optimal control of partially observable Markov processes*, Ph.D. Thesis, Stanford University.

Sonu, E., Chen, Y., and Doshi, P. (2017). Decision-theoretic planning under anonymity in agent populations, *Journal of Artificial Intelligence Research* **59**, pp. 725–770, doi:10.1613/jair.5449, https://doi.org/10.1613/jair.5449.

Sonu, E. and Doshi, P. (2013). Bimodal switching for online planning in multi-agent settings, in *International Joint Conference on Artificial Intelligence (IJCAI)*, pp. 360–366.

Spaan, M. and Vlassis, N. (2004). A point-based POMDP algorithm for robot planning, in *IEEE International Conference on Robotics and Automation (ICRA)*, pp. 2399–2404.

Spaan, M. T. and Melo, F. S. (2008). Interaction-driven Markov games for decentralized multiagent planning under uncertainty, in *Proceedings of the 7th International Joint Conference on Autonomous Agents and Multiagent Systems — Volume 1* (International Foundation for Autonomous Agents and Multiagent Systems), pp. 525–532.

Stone, P., Kaminka, G. A., Kraus, S., and Rosenschein, J. S. (2010). *Ad hoc autonomous agent teams: Collaboration without pre-coordination*, in *Proceedings of the 24th AAAI Conference on Artificial Intelligence, AAAI 2010*, Atlanta, Georgia, USA, pp. 1504–1509.

Stroupe, A. W. and Balch, T. (2011). Value-based observation with robot teams (vbort) using probabilistic techniques, in *Proceedings of International Conference on Advanced Robotics*.

Sutton, R. and Barto, A. (1998a). *Reinforcement Learning: An Introduction* (MIT Press, Cambridge, Massachusetts, USA).

Sutton, R. S. and Barto, A. G. (1998b). *Reinforcement Learning: An Introduction*, Vol. 1 (MIT press Cambridge).

Szabó, G. and Fáth, G. (2006). Evolutionary games on graphs, *Physics Reports* **446**, 46, pp. 97–216.

Tanner, H. G. and Jadbabaie, A. (2003). Stable flocking of mobile agents, *42nd IEEE International Conference on Decision and Control* (IEEE Cat. No. 03CH37475), **2**, pp. 2010–2015.

Taylor, G. W. and Hinton, G. E. (2009). Factored conditional restricted Boltzmann Machines for modeling motion style, in *Proceedings of the 26th Annual International Conference on Machine Learning*, ICML '09 (ACM), ISBN 978-1-60558-516-1, pp. 1025–1032, doi:10.1145/1553374.1553505, http://dx.doi.org/10.1145/1553374.1553505.

Thomson, B. and Young, S. (2010). Bayesian update of dialogue state: A POMDP framework for spoken dialogue systems, *Computer Speech & Language* **24**, 4, pp. 562–588.

Tiwari, R., Jain, P., Butail, S., Baliyarasimhuni, S. P., and Goodrich, M. A. (2017). Effect of leader placement on robotic swarm control, in *Proceedings of the 16th Conference on Autonomous Agents and MultiAgent Systems, AAMAS 2017*, São Paulo, Brazil, pp. 1387–1394.

Toner, J. and Tu, Y. (1998). Flocks, herds, and schools: A quantitative theory of flocking, *Physical Review E* **58**, 4, pp. 4828–4858.

Tosic, P. T. and Agha, G. A. (2004). Maximal clique based distributed coalition formation for task allocation in large-scale MultiAgent systems, in *Proceedings of International Conference on Massively MultiAgent Systems*, pp. 104–120.

Vasconcelos, W. W., Kollingbaum, M. J., and Norman, T. J. (2009). Normative conflict resolution in MultiAgent systems, *Autonomous Agents and Multi-Agent Systems* **19**, 2, pp. 124–152.

Vickrey, W. (1961). Counterspeculation, auctions, and competitive sealed tenders, *Journal of Finance* **16**, 1, pp. 8–37.

Vicsek, T., Czirok, A., Ben-Jacob, E., Cohen, I., Shochet, O. (1995). Novel type of phase transitions in a system of self-driven particles. *Physics Review Letter*, **75**(6), pp. 1226–1229.

Vidal, J. (2002). The effects of cooperation on multiagent search in task-oriented domains, in *Proceedings of 1st International Joint Conference on Autonomous Agents and Multiagent Systems*, pp. 453–454.

Vidal, J. (2003). A method for solving distributed service allocation problems, *Web Intelligence & Agent Systems* **1**, 2, pp. 139–146.

Villatoro, D., Sabater-Mir, J., and Sen, S. (2011). Social instruments for robust convention emergence, in *International Joint Conference on Artificial Intelligence*, Vol. 11, pp. 420–425.

Villatoro, D., Sabater-Mir, J., and Sen, S. (2013). Robust convention emergence in social networks through self-reinforcing structures dissolution, *ACM Transactions on Autonomous and Adaptive Systems (TAAS)* **8**, 1, p. 2.

Villatoro, D., Sen, S., and Sabater-Mir, J. (2009). Topology and memory effect on convention emergence, in *Proceedings of the 2009 IEEE/WIC/ACM*

International Joint Conference on Web Intelligence and Intelligent Agent Technology-Volume 02 (IEEE Computer Society), pp. 233–240.

Vogiatzis, G., MacGillivray, I., and Chli, M. (2010). A probabilistic model for trust and reputation, in *Proceedings of the 9th International Conference on Autonomous Agents and Multiagent Systems (AAMAS'10)*, pp. 225–232.

Wang, F. Y. and Lansing, S. J. (2004). From artificial life to artificial societies — new methods for studies of complex social systems, *Complex Systems & Complexity Science* **1**, 1, pp. 33–41.

Wang, M., Wang, H., Vogel, D., Kumar, K., and Chiu, D. K. (2009). Agent-based negotiation and decision making for dynamic supply chain formation, *Engineering Application of Artificial Intelligence* **22**, 7, pp. 1046–1055.

Wang, W. and Jiang, Y. (2014). Community-aware task allocation for social networked multiagent systems, *IEEE Transactions on Cybernetics* **44**, 9, pp. 1529–1543.

Wang, W. and Jiang, Y. (2015). Multiagent-based allocation of complex tasks in social networks, *IEEE Transactions on Emerging Topics in Computing* **3**, 4, pp. 1–1.

Wang, Y., Cai, Z., Yin, G., Gao, Y., Tong, X., and Wu, G. (2016). An incentive mechanism with privacy protection in mobile crowdsourcing systems, *Computer Networks* **102**, pp. 157–171.

Washington, R. (1997). BI-POMDP: Bounded, incremental partially observable Markov model planning, in *European Conference on Planning (ECP)*, pp. 440–451.

Watkins, C. J. C. H. (1989). Learning from delayed rewards, *Robotics & Autonomous Systems* **15**, 4, pp. 233–235.

Watkins, C. J. C. H. and Dayan, P. (1992). Technical note q-learning, *Machine Learning* **8**, pp. 279–292.

Weerdt, M. D., Zhang, Y., and Klos, T. (2007). Distributed task allocation in social networks, in *Proceedings of 6th International Joint Conference on Autonomous Agents and Multiagent Systems*, p. 76.

Weerdt, M. M. D., Zhang, Y., and Klos, T. (2012). Multiagent task allocation in social networks, *Autonomous Agents and MultiAgent Systems* **25**, 1, pp. 46–86.

Weibull, J. W. (1997). *Evolutionary Game Theory* (MIT press).

Weiss, G. (ed.) (2013). *Multiagent Systems, 2nd edition* (MIT Press, USA).

White, C. C., III (1991). A survey of solution techniques for the partially observed Markov decision process, *Annals of Operations Research* **32**, 1, pp. 215–230.

White III, C. C. and Cheong, T. (2012). In-transit perishable product inspection, *Transportation Research Part E: Logistics and Transportation Review* **48**, 1, pp. 310–330.

Wiering, M. and van Ottelo, M. (2012). *Reinforcement Learning: State of the Art* (Springer).

Wikle, T. and Bailey, G. (1997). The spatial diffusion of linguistic features in Oklahoma, *Proceedings of the Oklahoma Academy of Science*, **77**, pp. 1–15.

Williams, C., Robu, V., Gerding, E., and Jennings, N. (2011). Using Gaussian processes to optimise concession in complex negotiations against unknown opponents, in *Proceedings of IJCAI'2011* (AAAI Press), pp. 432–438.

Williams, C. R. (2012). *Practical strategies for agent-based negotiation in complex environments*, Ph.D. Thesis, University of Southampton.

Wolf, T. B. and Kochenderfer, M. J. (2011). Aircraft collision avoidance using Monte Carlo real-time belief space search, *Journal of Intelligent and Robotic Systems* **64**, 2, pp. 277–298.

Wooldridge, M. (2009). *An Introduction to Multiagent Systems* (John Wiley & Sons).

Wooldridge, M. and Jennings, N. R. (1995). Intelligent agents: Theory and practice, *The Knowledge Engineering Review* **10**, 02, pp. 115–152.

Wu, F., Zilberstein, S., and Chen, X. (2011a). Online planning for *ad hoc* autonomous agent teams, in *International Joint Conference on Artificial Intelligence (IJCAI)*, pp. 439–445.

Wu, F., Zilbersteinb, S., and Chen, X. (2011b). Online planning for Multi-Agent systems with bounded communication, *Artificial Intelligence* **175**, 2, pp. 487–511.

Wu, Z., Fang, H., and She, Y. (2012). Weighted average prediction for improving consensus performance of second-order delayed MultiAgent systems, *IEEE Trans. Systems, Man, and Cybernetics, Part B* **42**, 5, pp. 1501–1508.

Wunder, M., Littman, M. L., and Babes, M. (2010). Classes of multi-agent q-learning dynamics with epsilon-greedy exploration, in *Proceedings of the 27th International Conference on Machine Learning (ICML-10)*, pp. 1167–1174.

Xu, L. and Weigand, H. (2001). The evolution of the contract net protocol, in *Proceedings of International Conference on Advances in Web-Age Information Management*, pp. 257–266.

Xueguang, C. and Haigang, S. (2004). Further extensions of fipa contract net protocol: threshold plus doa, in *Proceedings of ACM Symposium on Applied Computing*, pp. 45–51.

Yan, F., Li, Z., and Jiang, Y. (2016). Controllable uncertain opinion diffusion under confidence bound and unpredicted diffusion probability, *Physica A Statistical Mechanics & Its Applications* **449**, pp. 85–100.

Yang, J., Zhang, H., Ling, Y., and Pan, C. (2014). Task allocation for wireless sensor network using modified binary particle swarm optimization, *IEEE Sensors Journal* **14**, 3, pp. 882–892.

Yang, T., Meng, Z., Hao, J., Sen, S., and Yu, C. (2016). Accelerating norm emergence through hierarchical heuristic learning, in *ECAI 2016 — 22nd European Conference on Artificial Intelligence, 29 August–2 September 2016, The Hague, The Netherlands — Including Prestigious Applications of Artificial Intelligence (PAIS 2016)*, pp. 1344–1352.

Ye, D., Zhang, M., and Sutanto, D. (2011). A hybrid multiagent framework with Q-learning for power grid systems restoration, *IEEE Transactions on Power Systems* **26**, 4, pp. 2434–2441.

Ye, D., Zhang, M., and Sutanto, D. (2013). Self-adaptation-based dynamic coalition formation in a distributed agent network: A mechanism and a brief survey, *IEEE Transactions on Parallel & Distributed Systems* **24**, 5, pp. 1042–1051.

Young, H. P. (1996). The economics of convention, *Journal of Economic Perspectives* **10**, 2, pp. 105–122.

Younis, M., Akkaya, K., and Kunjithapatham, A. (2003). Optimization of task allocation in a cluster-based sensor network, in *Proceedings of 8th IEEE International Symposium on Computers and Communication*, Vol. 1, pp. 329–334.

Yousefi, S., Weinreich, I., and Reinarz, D. (2005). Wavelet-based prediction of oil prices, *Chaos Solitons Fractals* **25**, 2, pp. 265–275.

Yu, C., Liu, R., and Wang, Y. (2016a). Social learning in networked agent societies, in *IEEE International Conference on Agents*, pp. 57–62.

Yu, C., Lv, H., Ren, F., Bao, H., and Hao, J. (2015a). Hierarchical learning for emergence of social norms in networked multiagent systems, in *AI 2015: Advances in Artificial Intelligence* (Springer), pp. 630–643.

Yu, C., Lv, H., Sen, S., Hao, J., Ren, F., and Liu, R. (2016b). An adaptive learning framework for efficient emergence of social norms, in *Proceedings of the 2016 International Conference on Autonomous Agents & Multiagent Systems* (International Foundation for Autonomous Agents and Multiagent Systems), pp. 1307–1308.

Yu, C., Lv, H., Sen, S., Ren, F., and Tan, G. (2016c). *Adaptive Learning for Efficient Emergence of Social Norms in Networked Multiagent Systems* (Springer International Publishing).

Yu, C., Tan, G., Lv, H., Wang, Z., Meng, J., Hao, J., and Ren, F. (2016d). Modelling adaptive learning behaviours for consensus formation in human societies, *Scientific reports* **6**.

Yu, C., Zhang, M., and Ren, F. (2014). Collective learning for the emergence of social norms in networked multiagent systems, *IEEE Transactions on Cybernetics* **44**, 12, pp. 2342–2355.

Yu, C., Zhang, M., Ren, F., and Luo, X. (2013a). Emergence of social norms through collective learning in networked agent societies, in *Proceedings of the 2013 International Conference on Autonomous Agents and MultiAgent Systems*, pp. 475–482.

Yu, H., Miao, C., An, B., Leung, C., and Lesser, V. R. (2013b). A reputation management approach for resource constrained trustee agents, in *Proceedings of the 23rd International Joint Conference on Artificial Intelligence (IJCAI'13)*, pp. 418–424.

Yu, H., Miao, C., Chen, Y., Fauvel, S., Li, X., and Lesser, V. R. (2017). Algorithmic management for improving collective productivity in crowdsourcing, *Scientific Reports* **7**, 12541, doi:10.1038/s41598-017-12757-x.

Yu, H., Miao, C., Leung, C., Chen, Y., Fauvel, S., Lesser, V. R., and Yang, Q. (2016e). Mitigating herding in hierarchical crowdsourcing networks, *Scientific Reports* **6**, 4, doi:10.1038/s41598-016-0011-6.

Yu, H., Miao, C., Shen, Z., Leung, C., Chen, Y., and Yang, Q. (2015b). Efficient task sub-delegation for crowdsourcing, in *Proceedings of the 29th AAAI Conference on Artificial Intelligence (AAAI-15)*, pp. 1305–1311.

Yu, H., Miao, C., Shen, Z., Lin, J., Leung, C., and Yang, Q. (2016f). Infusing human factors into algorithmic crowdsourcing, in *Proceedings of the 28th Conference on Innovative Applications of Artificial Intelligence (IAAI-16)*, pp. 4062–4063.

Yu, H., Shen, Z., Leung, C., Miao, C., and Lesser, V. R. (2013c). A survey of MultiAgent trust management systems, *IEEE Access* **1**, 1, pp. 35–50.

Yu, H., Shen, Z., Miao, C., and An, B. (2013d). A reputation-aware decision-making approach for improving the efficiency of crowdsourcing systems, in *Proceedings of the 12th International Conference on Autonomous Agents and MultiAgent Systems (AAMAS'13)*, pp. 1315–1316.

Yu, Y. and Prasanna, V. K. (2005). Energy-balanced task allocation for collaborative processing in wireless sensor networks, *Mobile Networks and Applications* **10**, 1, pp. 115–131.

Zeng, D. and Sycara, K. (1997). How can an agent learn to negotiate? in *Intelligent Agents III Agent Theories, Architectures, and Languages* (Springer), pp. 233–244.

Zeng, D. and Sycara, K. (1998). Bayesian learning in negotiation, *International Journal of Human Computer Studies* **48**, 1, pp. 125–141.

Zeng, Y., Chen, Y., and Doshi, P. (2011a). Approximating behavioral equivalence of models using top-k policy paths (extended abstract), in *International Conference on Autonomous Agents and multiagent Systems (AAMAS)*, pp. 1229–1230.

Zeng, Y. and Doshi, P. (2009). Speeding up exact solutions of interactive dynamic influence diagrams using action equivalence, in *IJCAI 2009, Proceedings of the 21st International Joint Conference on Artificial Intelligence, Pasadena, California*, USA, pp. 1996–2001.

Zeng, Y. and Doshi, P. (2012). Exploiting model equivalences for solving interactive dynamic influence diagrams, *Journal of Artificial Intelligence Research (JAIR)* **43**, pp. 211–255.

Zeng, Y., Doshi, P., Pan, Y., Mao, H., Chandrasekaran, M., and Luo, J. (2011b). Utilizing partial policies for identifying equivalence of behavioral models, in *25th Conference on Artificial Intelligence (AAAI)*, pp. 1083–1088.

Zeng, Y., Mao, H., Pan, Y., and Luo, J. (2012). Improved use of partial policies for identifying behavioral equivalence, in *Autonomous Agents and MultiAgent Systems Conference (AAMAS)*, pp. 1015–1022.

Zhang, C., Abdallah, S., and Lesser, V. (2009). Integrating organizational control into MultiAgent learning, in *Proceedings of the 8th International Conference on Autonomous Agents and Multiagent Systems*. Vol. 2, pp. 757–764.

Zhang, C., Lesser, V., and Abdallah, S. (2010). Self-organization for coordinating decentralized reinforcement learning, in *Proceedings of the 9th International*

Conference on Autonomous Agents and Multiagent Systems, Vol. 1, pp. 739–746.

Zhang, C. and Lesser, V. R. (2010). MultiAgent learning with policy prediction, in *Proceedings of the 24th AAAI Conference on Artificial Intelligence*, pp. 927–934.

Zhang, N. and Zhang, W. (2001). Speeding up the convergence of value iteration in partially observable Markov decision processes, *Journal of Artificial Intelligence Research* **14**, 1, pp. 29–51.

Zhang, Y. and Weerdt, M. D. (2007). VCG-based truthful mechanisms for social task allocation, *Proceedings of the Fifth European Workshop on MultiAgent Systems (EUMAS-07)*, pp. 378–394.

Zhang, Z. and Chen, X. (2012). FHHOP: A factored hybrid heuristic online planning algorithm for large POMDPs, in *Conference on Uncertainty in Artificial Intelligence (UAI)*, pp. 934–943.

Zhang, Z., Fu, Q., Zhang, X., and Liu, Q. (2016). Reasoning and predicting POMDP planning complexity via covering numbers, *Frontiers of Computer Science* **10**, 4, pp. 726–740.

Zhang, Z., Hsu, D., and Lee, W. (2014). Covering number for efficient heuristic-based POMDP planning, in *International Conference on Machine Learning (ICML)*, pp. 28–36.

Zhang, Z., Hsu, D., Lee, W., Lim, Z., and Bai, A. (2015). PLEASE: Palm leaf search for POMDPs with large observation spaces, in *International Conference on Automated Planning and Scheduling (ICAPS)*, pp. 249–257.

Zhang, Z., Littman, M., and Chen, X. (2012). Covering number as a complexity measure for POMDP planning and learning, in *AAAI Conference on Artificial Intelligence (AAAI)*, pp. 1853–1859.

Zhang, Z. and Liu, Q. (2016). Covering number: Analyses for approximate continuous-state POMDP planning, in *International Conference on Autonomous Agents and Multiagent Systems (AAMAS)*, pp. 1293–1294.

Zhao, D., Li, X. Y., and Ma, H. (2014). How to crowdsource tasks truthfully without sacrificing utility: Online incentive mechanisms with budget constraint, in *Proceedings of 33rd IEEE International Conference on Computer Communications*, pp. 1213–1221.

Zhao, D., Li, X. Y., and Ma, H. (2016a). Budget-feasible online incentive mechanisms for crowdsourcing tasks truthfully, *IEEE/ACM Transactions on Networking* **24**, 2, pp. 647–661.

Zhao, D., Ramchurn, S. D., and Jennings, N. R. (2016b). Fault tolerant mechanism design for general task allocation, in *Proceedings of 15th International Conference on Autonomous Agents and Multiagent Systems*, pp. 323–331.

Zheng, X. and Koenig, S. (2009). Negotiation with reaction functions for solving complex task allocation problems, in *Proceedings of IEEE/RSJ International Conference on Intelligent Robots and Systems*, pp. 4811–4816.

Zheng, X. and Koenig, S. (2011). Generalized reaction functions for solving complex-task allocation problems, in *Proceedings of 22nd International Joint Conference on Artificial Intelligence*, pp. 478–483.

Zhu, Y., Zhang, Q., Zhu, H., Yu, J., Cao, J., and Ni, L. M. (2014). Towards truthful mechanisms for mobile crowdsourcing with dynamic smartphones, in *Proceedings of IEEE International Conference on Distributed Computing Systems*, pp. 11–20.

Zinkevich, M. (2003). Online convex programming and generalized infinitesimal gradient ascent, *Proceedings of the 20th International Conference on Machine Learning*, pp. 928–936.

Zlot, R., Stentz, A., Dias, M. B., and Thayer, S. (2002). Multi-robot exploration controlled by a market economy, in *Proceedings of IEEE International Conference on Robotics and Automation*, pp. 3016–3023.

Zlot, R. M. (2006). An auction-based approach to complex task allocation for multirobot teams, *Proceedings of the IEEE* **94**, 7, pp. 1257–1270.

Index

295

Printed in the United States
by Bookmasters

Printed in the United States
By Bookmasters